Victor Andreevich Toponogov

with the editorial assistance of
Vladimir Y. Rovenski

Differential Geometry of Curves and Surfaces

A Concise Guide

Birkhäuser
Boston • Basel • Berlin

Victor A. Toponogov *(deceased)*
Department of Analysis and Geometry
Sobolev Institute of Mathematics
Siberian Branch of the Russian Academy
   of Sciences
Novosibirsk-90, 630090
Russia

With the editorial assistance of:
Vladimir Y. Rovenski
Department of Mathematics
University of Haifa
Haifa, Israel

Cover design by Alex Gerasev.

AMS Subject Classification: 53-01, 53Axx, 53A04, 53A05, 53A55, 53B20, 53B21, 53C20, 53C21

Library of Congress Control Number: 2005048111

ISBN-10 0-8176-4384-2   eISBN 0-8176-4402-4
ISBN-13 978-0-8176-4384-3

Printed on acid-free paper.

©2006 Birkhäuser Boston   *Birkhäuser*

Printed in the United States of America.   (TXQ/EB)

9 8 7 6 5 4 3 2 1

www.birkhauser.com

Contents

Preface

This concise guide to the differential geometry of curves and surfaces can be recommended to first-year graduate students, strong senior students, and students specializing in geometry. The material is given in two parallel streams.

The first stream contains the standard theoretical material on differential geometry of curves and surfaces. It contains a small number of exercises and simple problems of a local nature. It includes the whole of Chapter 1 except for the problems (Sections 1.5, 1.7, 1.10) and Section 1.11, about the phase length of a curve, and the whole of Chapter 2 except for Section 2.6, about classes of surfaces, Theorems 2.8.1–2.8.4, the problems (Sections 2.7.4, 2.8.3) and the appendix (Section 2.9).

The second stream contains more difficult and additional material and formulations of some complicated but important theorems, for example, a proof of A.D. Aleksandrov's comparison theorem about the angles of a triangle on a convex surface,[1] formulations of A.V. Pogorelov's theorem about rigidity of convex surfaces, and S.N. Bernstein's theorem about saddle surfaces. In the last case, the formulations are discussed in detail.

A distinctive feature of the book is a large collection (80 to 90) of *nonstandard and original problems* that introduce the student into the real world of geometry. Most of these problems are new and are not to be found in other textbooks or books of problems. The solutions to them require inventiveness and geometrical intuition. In this respect, this book is not far from W. Blaschke's well-known

---

[1] A generalization of Aleksandrov's global angle comparison theorem to Riemannian spaces of arbitrary dimension is known as Toponogov's theorem.

manuscript [Bl], but it contains a number of problems more contemporary in theme. The key to these problems is the notion of *curvature*: the curvature of a curve, principal curvatures, and the Gaussian curvature of a surface. Almost all the problems are given with their solutions, although the hope of the author is that an honest student will solve them without assistance, and only in exceptional cases will look at the text for a solution. Since the problems are given in increasing order of difficulty, even the most difficult of them should be solvable by a motivated reader. In some cases, only short instructions are given. In the author's opinion, it is the large number of original problems that makes this textbook interesting and useful.

Chapter 3, *Intrinsic Geometry of a Surface*, starts from the main notion of a *covariant derivative of a vector field along a curve*. The definition is based on extrinsic geometrical properties of a surface. Then it is proven that the covariant derivative of a vector field is an object of the intrinsic geometry of a surface, and the later training material is not related to an extrinsic geometry. So Chapter 3 can be considered an *introduction to n-dimensional Riemannian geometry that keeps the simplicity and clarity of the 2-dimensional case.*

The main theorems about geodesics and shortest paths are proven by methods that can be easily extended to n-dimensional situations almost without alteration. The Aleksandrov comparison theorem, Theorem 3.9.1, for the angles of a triangle is the high point in Chapter 3.

The author is one of the founders of *CAT(k)-spaces theory,*[2] where the *comparison theorem for the angles of a triangle*, or more exactly its generalization by the author to multidimensional Riemannian manifolds, takes the place of the basic property of CAT(k)-spaces.

Acknowledgments. The author gratefully thanks his student and colleagues who have contributed to this volume. Essential help was given by E.D. Rodionov, V.V. Slavski, V.Yu. Rovenski, V.V. Ivanov, V.A. Sharafutdinov, and V.K. Ionin.

---

[2] The initials are in honor of E. Cartan, A.D. Aleksandrov, and V.A. Toponogov.

About the Author

Professor Victor Andreevich Toponogov, a well-known Russian geometer, was born on March 6, 1930, and grew up in the city of Tomsk, in Russia. During Toponogov's childhood, his father was subjected to Soviet repression. After finishing school in 1948, Toponogov entered the Department of Mechanics and Mathematics at Tomsk University, and graduated in 1953 with honors.

In spite of an active social position and receiving high marks in his studies, the stamp of "son of an enemy of the people" left Toponogov with little hope of continuing his education at the postgraduate level. However, after Joseph Stalin's death in March 1953, the situation in the USSR changed, and Toponogov became a postgraduate student at Tomsk University. Toponogov's scientific interests were influenced by his scientific advisor, Professor A.I. Fet (a recognized topologist and specialist in variational calculus in the large, a pupil of L.A. Lusternik) and by the works of Academician A.D. Aleksandrov.[1]

In 1956, V.A. Toponogov moved to Novosibirsk, where in April 1957 he became a research scientist at the Institute of Radio-Physics and Electronics, then directed by the well-known physicist Y.B. Rumer. In December 1958, Toponogov defended his Ph.D. thesis at Moscow State University. In his dissertation, the Aleksandrov convexity condition was extended to multidimensional Riemannian manifolds. Later, this theorem came to be called the *Toponogov (comparison) theorem*.[2] In April 1961, Toponogov moved to the Institute of Mathematics and

---

[1] Aleksandr Danilovich Aleksandrov (1912–1999).

[2] Meyer, W.T. *Toponogov's Theorem and Applications*. Lecture Notes, College on Differential Geometry, Trieste. 1989.

Computer Center of the Siberian Branch of the Russian Academy of Sciences at its inception. All his subsequent scientific activity is related to the Institute of Mathematics. In 1968, at this institute he defended his doctoral thesis on the theme *"Extremal problems for Riemannian spaces with curvature bounded from above."*

From 1980 to 1982, Toponogov was deputy director of the Institute of Mathematics, and from 1982 to 2000 he was head of one of the laboratories of the institute. In 2001 he became Chief Scientist of the Department of Analysis and Geometry.

The first thirty years of Toponogov's scientific life were devoted to one of the most important divisions of modern geometry: *Riemannian geometry in the large.*

From secondary-school mathematics, everybody has learned something about synthetic methods in geometry, concerned with triangles, conditions of their equality and similarity, etc. From the Archimedean era, analytical methods have come to penetrate geometry: this is expressed most completely in the theory of surfaces, created by Gauss. Since that time, these methods have played a leading part in differential geometry. In the fundamental works of A.D. Aleksandrov, synthetic methods are again used, because the objects under study are not smooth enough for applications of the methods of classical analysis. In the creative work of V.A. Toponogov, both of these methods, synthetic and analytic, are in harmonic correlation.

The classic result in this area is the Toponogov theorem about the angles of a triangle composed of geodesics. This in-depth theorem is the basis of modern investigations of the relations between curvature properties, geodesic behavior, and the topological structure of Riemannian spaces. In the proof of this theorem, some ideas of A.D. Aleksandrov are combined with the in-depth analytical technique related to the Jacobi differential equation.

The methods developed by V.A. Toponogov allowed him to obtain a sequence of fundamental results such as characteristics of the multidimensional sphere by estimates of the Riemannian curvature and diameter, the solution to the Rauch problem for the even-dimensional case, and the theorem about the structure of Riemannian space with nonnegative curvature containing a straight line (i.e., the shortest path that may be limitlessly extended in both directions). This and other theorems of V.A. Toponogov are included in monographs and textbooks written by a number of authors. His methods have had a great influence on modern Riemannian geometry.

During the last fifteen years of his life, V.A. Toponogov devoted himself to differential geometry of two-dimensional surfaces in three-dimensional Euclidean space. He made essential progress in a direction related to the *Efimov theorem* about the nonexistence of isometric embedding of a complete Riemannian metric with a separated-from-zero negative curvature into three-dimensional Euclidean space, and with the *Milnor conjecture* declaring that an embedding with a sum of absolute values of principal curvatures uniformly separated from zero does not exist.

Toponogov devoted much effort to the training of young mathematicians. He was a lecturer at Novosibirsk State University and Novosibirsk State Pedagogical University for more than forty-five years. More than ten of his pupils defended their Ph.D. theses, and seven their doctoral degrees.

V.A. Toponogov passed away on November 21, 2004 and is survived by his wife, Ljudmila Pavlovna Goncharova, and three sons.

*Differential Geometry of
Curves and Surfaces*

1

Theory of Curves in Three-dimensional Euclidean Space and in the Plane

1.1 Preliminaries

An example of a vector space is $\mathbb{R}^n$, the set of n-tuples $(x_1, \ldots, x_n)$ of real numbers. Three vectors $\vec{i} = (1, 0, 0)$, $\vec{j} = (0, 1, 0)$, $\vec{k} = (0, 0, 1)$ form a basis of the space $\mathbb{R}^3$. A *ball* in $\mathbb{R}^n$ with center $P(x_1^0, \ldots, x_n^0)$ and radius $\varepsilon > 0$ is the set $B(P, \varepsilon) = \{(x_1, \ldots, x_n) \in \mathbb{R}^n : \sum_{i=1}^{n} (x_i - x_i^0)^2 < \varepsilon^2\}$. A set $U \subset \mathbb{R}^n$ is *open* if for each $P \in \mathbb{R}^n$ there is a ball $B(P, \varepsilon) \subset U$.

Definition 1.1.1. If $\vec{a} = a_1\vec{i} + a_2\vec{j} + a_3\vec{k}$ and $\vec{b} = b_1\vec{i} + b_2\vec{j} + b_3\vec{k}$ are vectors in $\mathbb{R}^3$, then their *scalar product* $\langle \vec{a}, \vec{b} \rangle$ and *vector product* $\vec{a} \times \vec{b}$ are

$$\langle \vec{a}, \vec{b} \rangle = a_1 b_1 + a_2 b_2 + a_3 b_3, \qquad \vec{a} \times \vec{b} = \det \begin{pmatrix} \vec{i} & \vec{j} & \vec{k} \\ a_1 & a_2 & a_3 \\ b_1 & b_2 & b_3 \end{pmatrix}.$$

The *triple product* of vectors $\vec{a}$, $\vec{b}$, and $\vec{c} = c_1\vec{i} + c_2\vec{j} + c_3\vec{k}$ is

$$(\vec{a} \cdot \vec{b} \cdot \vec{c}) = \det \begin{pmatrix} a_1 & a_2 & a_3 \\ b_1 & b_2 & b_3 \\ c_1 & c_2 & c_3 \end{pmatrix}.$$

Definition 1.1.2. A *linear transformation* is a function $T : V \to W$ of vector spaces such that $T(\lambda \vec{a} + \mu \vec{b}) = \lambda T(\vec{a}) + \mu T(\vec{b})$ for all $\lambda, \mu \in \mathbb{R}$ and $\vec{a}, \vec{b} \in V$. An *isomorphism* is a one-to-one linear transformation. A real number λ is an *eigenvalue* of a linear transformation $T : V \to V$ if there is a nonzero vector $\vec{a}$ (called an *eigenvector*) such that $T(\vec{a}) = \lambda \vec{a}$.

Definition 1.1.3. If a map $\varphi: M \to N$ is continuous and bijective, and if its inverse map $\psi = \varphi^{-1}: N \to M$ is also continuous, then φ is a *homeomorphism* and M and N are said to be *homeomorphic*. The *Jacobi matrix* of a differentiable map $\varphi: \mathbb{R}^n \to \mathbb{R}^m$ is

$$J = \begin{pmatrix} \frac{\partial f_1}{\partial x_1} & \cdots & \frac{\partial f_1}{\partial x_n} \\ \vdots & \ddots & \vdots \\ \frac{\partial f_m}{\partial x_1} & \cdots & \frac{\partial f_m}{\partial x_n} \end{pmatrix}.$$

A differentiable map $\varphi: M \to N$ is a *diffeomorphism* if there is a differentiable map $\psi: N \to M$ such that $\varphi \circ \psi = I$ (where I is the identity map) and $\psi \circ \varphi = I$.

Theorem 1.1.1 (Inverse function theorem). *Let $U \subset \mathbb{R}^n$ be an open set, $P \in U$, and $\varphi: U \to \mathbb{R}^n$. If $\det J(P) \neq 0$, then there exist neighborhoods V_P of P and $V_{\varphi(P)}$ of $\varphi(P)$ such that $\varphi|_{V_P}: V_P \to V_{\varphi(P)}$ is a diffeomorphism.*

For $y = (y_1, \ldots, y_n)$ and fixed integer $i \in [1, n]$, set $\tilde{y} = (y_1, \ldots, y_{i-1}, y_{i+1}, \ldots, y_n)$. If $W \subset \mathbb{R}^{n+1}$, then $\tilde{W} = \{\tilde{w}: w \in W\} \subset \mathbb{R}^n$ is a projection along the ith coordinate axes.

Theorem 1.1.2 (Implicit function theorem). *Let $\varphi: \mathbb{R}^{n+1} \to \mathbb{R}$ be a C^k $(k \geq 1)$ function, $P \in \mathbb{R}^{n+1}$, and $(\partial\varphi/\partial x_i)(P) \neq 0$ for some fixed i. Then there is a neighborhood W of P in $\mathbb{R}^{n+1}$ and a C^k function $f: \tilde{W} \to \mathbb{R}$ such that for $y = (y_1, \ldots, y_{n+1}) \in \mathbb{R}^{n+1}$, $f(y_1, \ldots, y_{n+1}) = 0$ if and only if $y_i = f(\tilde{y})$.*

Theorem 1.1.3 (Existence and uniqueness solution). *Let a map $\mathbf{f}: \mathbb{R}^{n+1} \to \mathbb{R}^n$ be continuous in a region $D = \{\|\vec{x} - \vec{x}_0\| \leq b, |t - t_0| \leq a\}$ and have bounded partial derivatives with respect to the coordinates of $\vec{x} \in \mathbb{R}^n$. Let $M = \sup \|\mathbf{f}(\vec{x}, t)\|$ over D. Then the differential equation $d\vec{x}/dt = \mathbf{f}(\vec{x}, t)$ has a unique solution on the interval $|t - t_0| \leq \min(T, b/M)$ satisfying $\vec{x}(t_0) = \vec{x}_0$.*

1.2 Definition and Methods of Presentation of Curves

We assume that a rectangular Cartesian coordinate system $(O; x, y, z)$ in three-dimensional Euclidean space $\mathbb{R}^3$ has been introduced.

Definition 1.2.1. A connected set γ in the space $\mathbb{R}^3$ (in the plane $\mathbb{R}^2$) is a *regular k-fold continuously differentiable curve* if there is a homeomorphism $\varphi: G \to \gamma$, where G is a line segment $[a, b]$ or a circle of radius 1, satisfying the following conditions:

  (1) $\varphi \in C^k$ $(k \geq 1)$,   (2) the rank of φ is maximal (equal to 1).

For $k = 1$ a curve γ is said to be *smooth*. Note that a regular curve γ of class C^k $(k \geq 1)$ is diffeomorphic either to a line segment or to a circle. Since a rectangular Cartesian coordinate system x, y, z is given in the space $\mathbb{R}^3$, a map φ is determined by a choice of the functions $x(t), y(t), z(t)$, where $t \in [a, b]$. The condition (1) means that these functions belong to class C^k, and the condition (2) means that the derivatives $x'(t)$, $y'(t)$, $z'(t)$ cannot simultaneously be zero for any t.

Any regular curve in $\mathbb{R}^3$ ($\mathbb{R}^2$) may be determined by one map $\varphi\colon x = x(t)$, $y = y(t)$, $z = z(t)$, where $t \in [a, b]$, and the equations $x = x(t)$, $y = y(t)$, $z = z(t)$ are called *parametric equations* of a curve γ. In the case that a regular curve is diffeomorphic to a circle, the functions $x(t)$, $y(t)$, $z(t)$ are periodic on $\mathbb{R}$ with period $b - a$, and the curve itself is called a *closed curve*. If φ is bijective, φ is called *simple*.

The Jordan curve theorem says that *a simple closed plane curve has an interior and an exterior*.

It is often convenient to use the *vector form* of parametric equations of a curve: $\vec{r} = \vec{r}(t) = x(t)\vec{i} + y(t)\vec{j} + z(t)\vec{k}$, where $\vec{i}$, $\vec{j}$, $\vec{k}$ are unit vectors of the axes OX, OY, OZ. If γ is a plane curve, then suppose $z(t) \equiv 0$.

The same curve (image) γ can be given by different parameterizations:

$$\vec{r} = \vec{r}_1(t) = x_1(t)\vec{i} + y_1(t)\vec{j} + z_1(t)\vec{k}, \qquad t \in (a, b),$$
$$\vec{r} = \vec{r}_2(\tau) = x_2(\tau)\vec{i} + y_2(\tau)\vec{j} + z_2(\tau)\vec{k}, \qquad \tau \in (c, d).$$

Then these vector functions $\vec{r}_1(t)$ and $\vec{r}_2(\tau)$ are related by a strictly monotonic transformation of parameters $t = t(\tau)\colon (c, d) \to (a, b)$ such that

(1) $\vec{r}_1(t(\tau)) = \vec{r}_2(\tau)$,

(2) $t'(\tau) \neq 0$ for all $\tau \in (c, d)$.

The existence of a function $t = t(\tau)$, its differentiability, and strong monotonic character follow from the definition of a regular curve and from the inverse function theorem.

Example 1.2.1. The parameterized regular space curve $x = a\cos t$, $y = a\sin t$, $z = bt$ lies on a cylinder $x^2 + y^2 = a^2$ and is called a *(right circular) helix* of pitch $2\pi b$ (Figure 2.17b). Here the parameter t measures the angle between the OX axis and the line joining the origin to the projection of the point $\vec{r}(t)$ over the XOY plane.

The parameterized space curve $x = at\cos t$, $y = at\sin t$, $z = bt$ lies on a cone $b^2(x^2 + y^2) = a^2 z^2$ and is called a *(circular) conic helix*.

Definition 1.2.2. A continuous curve γ is called *piecewise smooth (piecewise regular)* if there exist a finite number of points P_i ($i = 1, \ldots, k$) on γ such that each connected component of the set $\gamma \setminus \bigcup_i P_i$ is a smooth (regular) curve.

Example 1.2.2. The trajectory of a point on a circle of radius R rolling (without sliding) in the plane along another circle of radius R' is called a *cycloidal curve*. If the circle moves along and inside of a fixed circle, then the curve is a *hypocycloid*; if outside, then the curve is an *epicycloid*. Parameterizations of these plane curves are

$$x = R(m + 1)\cos(mt) - Rm\cos(mt + t),$$
$$y = R(m + 1)\sin(mt) - Rm\sin(mt + t) \tag{1.1}$$

where $m = R/R'$ is the *modulus*. For $m > 0$ we have *epicycloids*, for $m < 0$, *hypocycloids*.

All cycloidal curves are piecewise regular. They are closed (periodic) for m rational only. A *cardioid* is an epicycloid with modulus $m = 1$; it has one singular point. An *astroid* is a hypocycloid with modulus $m = -\frac{1}{4}$, see also Exercise 1.12.19. It has four singular points.

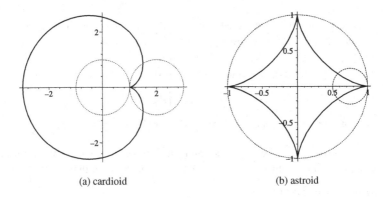

(a) cardioid   (b) astroid

Figure 1.1. Cycloidal curves.

Besides the *parametric presentation* of a curve γ in $\mathbb{R}^3$ ($\mathbb{R}^2$) there also exist other presentations.

Explicitly given curve.   A particular case of the parametric presentation of a curve is an *explicit presentation* of a curve, when the part of a parameter t is played by either the variable x, y, or z; i.e., either $x = x$, $y = f_1(x)$, $z = f_2(x)$; $x = f_1(y)$, $y = y$, $z = f_2(y)$; or $x = f_1(z)$, $y = f_2(z)$, $z = z$. An explicit presentation is especially convenient for a plane curve. In this case a curve coincides with a *graph* of some function f, and then the equation of the curve may be written either in the form $y = f(x)$ or $x = f(y)$.

Example 1.2.3. A *tractrix* (see Figure 2.12 a) can be presented as a graph $x = a\ln\frac{a - \sqrt{a^2 - y^2}}{y} + \sqrt{a^2 - y^2}$, $0 < y \le a$. It has one singular point $P(a, 0)$. For a parameterization of this plane curve see Exercise 1.12.22.

Implicitly given curve.   Let a differentiable map be given by

$$\mathbf{f}: \mathbb{R}^3 \to \mathbb{R}^2, \qquad \mathbf{f} = [f_1(x, y, z),\ f_2(x, y, z)].$$

Then from the implicit function theorem it follows that if $(0, 0)$ is a regular value of the map $\mathbf{f}$, then each connected component of the set $T = \mathbf{f}^{-1}(0, 0)$ is a smooth regular curve in $\mathbb{R}^3$. In other words, under the above given conditions a set of points in $\mathbb{R}^3$ whose coordinates satisfy the system of equations

$$f_1(x, y, z) = 0, \qquad f_2(x, y, z) = 0, \tag{1.2}$$

forms a smooth regular curve (more exactly, a finite number of smooth regular curves). This method is called an *implicit presentation of a curve*, and the system (1.2) is called the *implicit equations of a curve*. In the plane case, an implicit presentation of a curve is based on a function $f : \mathbb{R}^2 \to \mathbb{R}$ with the property that 0 is a regular value.

Recall that the value $(0, 0)$ of a map $\mathbf{f} = (f_1, f_2) \colon \mathbb{R}^3 \to \mathbb{R}^2$ is *regular* if the rank of the Jacobi matrix

$$J = \begin{pmatrix} \frac{\partial f_1}{\partial x} & \frac{\partial f_1}{\partial y} & \frac{\partial f_1}{\partial z} \\ \frac{\partial f_2}{\partial x} & \frac{\partial f_2}{\partial y} & \frac{\partial f_2}{\partial z} \end{pmatrix}$$

is 2 (or $\det J \neq 0$) at every point of the solution set of (1.2) .

Obviously, an explicit presentation of a curve is at the same time a parametric presentation, where the role of a parameter t is played by the x-coordinate, say. Conversely, if a regular curve is given by parametric equations, then in some neighborhood of an arbitrary point, as follows from the converse function theorem, there an its explicit presentation. Analogously, if a curve is presented by implicit equations, then in some neighborhood of an arbitrary point it admits an explicit presentation. The last statement can be deduced from the implicit function theorem.

Example 1.2.4. (a) The intersection of a sphere $x^2 + y^2 + z^2 = R^2$ of radius R with a cylinder $x^2 + y^2 = Rx$ of radius $\frac{R}{2}$ is a *Viviani curve* with one point of

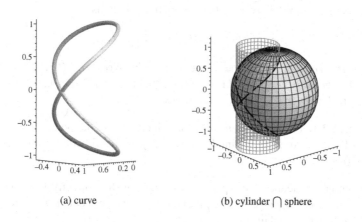

(a) curve   (b) cylinder $\bigcap$ sphere

Figure 1.2. Viviani window.

self-intersection. One can verify that $\vec{r} = [R \cos^2 t,\ R \cos t \sin t,\ R \sin t]$, $0 \leq t \leq 2\pi$, is a regular parameterization of the curve.

(b) The intersection of two cylinders with orthogonal axes, $x^2 + z^2 = R_1^2$ and $y^2 + z^2 = R_2^2$, of radii $R_1 \geq R_2$ is a *bicylinder* curve. One can verify that for $R_1 = R_2$ it degenerates to a pair of ellipses, and that

$$\vec{r} = \left[R_1 \cos t, \pm \sqrt{R_1^2 - R_2^2 \sin^2 t}, \; R_1 \sin t \right], \quad 0 \leq t \leq 2\pi,$$

is a regular parameterization of the two curve components.

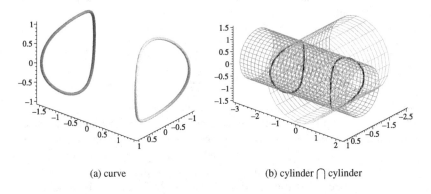

(a) curve   (b) cylinder $\bigcap$ cylinder

Figure 1.3. Bicylinder curve.

1.3 Tangent Line and Osculating Plane

Let a smooth curve γ be given by the parametric equations

$$\vec{r} = \vec{r}(t) = x(t)\vec{i} + y(t)\vec{j} + z(t)\vec{k}.$$

The *velocity vector* of $\vec{r}(t)$ at $t = t_0$ is the derivative $\vec{r}'(t_0) = x'(t_0)\vec{i} + y'(t_0)\vec{j} + z'(t_0)\vec{k}$. The *velocity vector field* is the vector function $\vec{r}'(t)$. The *speed* of $\vec{r}(t)$ at $t = t_0$ is the length $|\vec{r}'(t_0)|$ of the velocity vector.

Definition 1.3.1. The *tangent line* to a smooth curve γ at the point $P = \vec{r}(t_0)$ is the straight line through the point $P = \vec{r}(t_0) \in \gamma$ in the direction of the velocity vector $\vec{r}'(t_0)$.

One can easily deduce the equations of a tangent line directly from its definition. In the case of *parametric equations* of a curve we obtain $\vec{r} = \vec{R}(u) = \vec{r}(t_0) + u\vec{r}'(t_0)$, or in detail,

$$\begin{cases} x = x(t_0) + ux'(t_0), \\ y = y(t_0) + uy'(t_0), \\ z = z(t_0) + uz'(t_0), \end{cases} \tag{1.3}$$

or in canonical form,

$$\frac{x - x(t_0)}{x'(t_0)} = \frac{y - y(t_0)}{y'(t_0)} = \frac{z - z(t_0)}{z'(t_0)}. \tag{1.4}$$

In the case of *explicit equations* of a curve, $y = \varphi_1(x)$, $z = \varphi_2(x)$, the tangent line is given by the following equations:

$$x - x_0 = \frac{y - \varphi_1(x_0)}{\varphi_1'(x_0)} = \frac{z - \varphi_2(x_0)}{\varphi_2'(x_0)}. \tag{1.5}$$

Finally, if a curve γ is given by *implicit equations*

$$f_1(x, y, z) = 0, \qquad f_2(x, y, z) = 0,$$

and $P(x_0, y_0, z_0)$ belongs to γ, then the rank of the Jacobi matrix

$$J = \begin{pmatrix} \frac{\partial f_1}{\partial x} & \frac{\partial f_1}{\partial y} & \frac{\partial f_1}{\partial z} \\ \frac{\partial f_2}{\partial x} & \frac{\partial f_2}{\partial y} & \frac{\partial f_2}{\partial z} \end{pmatrix}$$

is 2 at P (i.e., rows and columns of J are each linearly independent). Assume for definiteness that the determinant

$$\begin{vmatrix} \frac{\partial f_1}{\partial x} & \frac{\partial f_1}{\partial y} \\ \frac{\partial f_2}{\partial x} & \frac{\partial f_2}{\partial y} \end{vmatrix}$$

is nonzero. Then by the implicit function theorem there exist a real number $\varepsilon > 0$ and differentiable functions $\varphi_1(x)$, $\varphi_2(y)$ such that for $|x - x_0| < \varepsilon$,

$$f_1(x, \varphi_1(x), \varphi_2(x)) \equiv 0, \qquad f_2(x, \varphi_1(x), \varphi_2(x)) \equiv 0.$$

Hence the equations of a tangent line to a curve γ at the point $P(x_0, y_0, z_0)$ are presented by (1.5), where the numbers $\varphi_1'(x_0)$ and $\varphi_2'(x_0)$ are solutions of the system of equations

$$\begin{cases} \frac{\partial f_1}{\partial x} + \frac{\partial f_1}{\partial y} \cdot \varphi_1'(x_0) + \frac{\partial f_1}{\partial z} \cdot \varphi_2'(x_0) = 0, \\ \frac{\partial f_2}{\partial x} + \frac{\partial f_2}{\partial y} \cdot \varphi_1'(x_0) + \frac{\partial f_2}{\partial z} \cdot \varphi_2'(x_0) = 0. \end{cases} \tag{1.6}$$

In the case of an implicit presentation of a plane curve $\gamma : f(x, y) = 0$, the equation of its tangent line can be written in the form

$$(\partial f / \partial x)(x_0, y_0)(x - x_0) + (\partial f / \partial y)(x_0, y_0)(y - y_0) = 0. \tag{1.7}$$

1.3.1 Geometric Characterization of a Tangent Line

Denote by d the length of a chord of a curve joining the points $P = \gamma(t_0)$ and $P_1 = \gamma(t_1)$, and by h the length of a perpendicular dropped from P_1 onto the tangent line to γ at the point P.

Theorem 1.3.1. $$\lim_{d\to 0}\frac{h}{d} = \lim_{t_1\to t_0}\frac{h}{d} = 0.$$

Proof. From the definition of magnitudes d and h one may deduce their expressions

$$d = |\vec{r}(t_1) - \vec{r}(t_0)|, \qquad h = \frac{|\vec{r}'(t_0) \times (\vec{r}(t_1) - \vec{r}(t_0))|}{|\vec{r}'(t_0)|}.$$

Then

$$\lim_{d\to 0}\frac{h}{d} = \lim_{t_1\to t_0}\frac{|\vec{r}'(t_0) \times (\vec{r}(t_1) - \vec{r}(t_0)|}{|\vec{r}'(t_0)| \cdot |\vec{r}(t_1) - \vec{r}(t_0)|}$$

$$= \lim_{t_1\to t_0}\frac{|\vec{r}'(t_0) \times \frac{\vec{r}(t_1)-\vec{r}(t_0)}{t_1-t_0}|}{|\vec{r}'(t_0)| \cdot |\frac{\vec{r}(t_1)-\vec{r}(t_0)}{t_1-t_0}|} = \frac{|\vec{r}'(t_1) \times \vec{r}'(t_0)|}{|\vec{r}'(t_0)|^2} = 0. \qquad \square$$

Theorem 1.3.1 explains the geometric characterization of a tangent line.

First of all, the theorem shows us that the tangent line l to a curve γ at the point $P = \gamma(t_0)$ is the limit of *secants* to γ that pass through P and an arbitrary point $P_1 = \gamma(t_1)$ for $t_1 \to t_0$. In fact, if we denote by α an angle between l and a secant PP_1, then $\frac{h}{d} = \sin\alpha$, and from Theorem 1.3.1 it follows that $\sin\alpha \to 0$ for $t_1 \to t_0$. From this our statement follows.

Secondly, Theorem 1.3.1 estimates an error that we obtain from replacing a curve γ by its tangent line l. Let $B_P(d) = \{x \in \mathbb{R}^3 \colon |x - P| < d\}$ be a ball with center P and radius d. Replace an arc $\gamma \cap B_P(d)$ of a curve γ by the line segment of l that belongs to $B_P(d)$. Then Theorem 1.3.1 claims that under such a change we make an error of higher order than the radius d of a ball. Also, this theorem allows us to give a geometric definition of a tangent line to a curve.

Denote by $\vec{\tau}(t_0)$ a unit vector that is parallel to $\vec{r}'(t_0)$: $\vec{\tau}(t_0) = \frac{\vec{r}'(t_0)}{|\vec{r}'(t_0)|}$. A straight line through the point $P = \gamma(t_0)$ that is orthogonal to the tangent line is called a *normal line*.

1.3.2 Osculating Plane

It is convenient to give a geometric definition of the osculating plane. Let a plane α with a unit normal $\vec{\beta}$ pass through a point $P = \vec{r}(t_0)$ of a curve γ. Denote by d the length of the chord of γ joining the points $P_0 = \vec{r}(t_0)$ and $P_1 = \vec{r}(t_1)$, and by h the length of the perpendicular dropped from P_1 onto the plane α.

Definition 1.3.2. A plane α is called an *osculating plane* to a curve γ at a point $P = \vec{r}(t_0)$ if

$$\lim_{d\to 0}\frac{h}{d^2} = \lim_{t_1\to t_0}\frac{h}{d^2} = 0.$$

Theorem 1.3.2. *At each point $P = \vec{r}(t_0)$ of a regular curve γ of class C^k ($k \geq 2$) there is an osculating plane α, and the vectors $\vec{r}'(t_0)$ and $\vec{r}''(t_0)$ are orthogonal to its normal vector $\vec{\beta}$.*

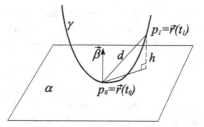

Figure 1.4. Osculating plane.

Proof. First, we shall prove the second statement, assuming the existence of an osculating plane to γ at a point $P = \vec{r}(t_0)$. From the definition of the magnitudes d and h it follows that

$$d = |\vec{r}(t_1) - \vec{r}(t_0)|, \qquad h = |\langle \vec{r}(t_1) - \vec{r}(t_0), \vec{\beta} \rangle|.$$

By Taylor's formula, $\vec{r}(t_1) - \vec{r}(t_0) = \vec{r}'(t_0)(t_1 - t_0) + \frac{1}{2}\vec{r}''(t_0)(t_1 - t_0)^2 + \vec{o}(|t_1 - t_0|^2)$. Hence

$$\lim_{d \to 0} \frac{h}{d^2} = \lim_{t_1 \to t_0} \frac{|\langle \vec{r}'(t_0)(t_1 - t_0) + \frac{1}{2}\vec{r}''(t_0)(t_1 - t_0)^2 + \vec{o}((t_1 - t_0)^2), \vec{\beta} \rangle|}{|\vec{r}(t_1) - \vec{r}(t_0)|^2}$$

$$= \lim_{t_1 \to t_0} \frac{\left| \frac{\langle \vec{r}'(t_0), \vec{\beta} \rangle}{t_1 - t_0} + \frac{1}{2}\langle \vec{r}''(t_0), \vec{\beta} \rangle + \frac{\langle \vec{o}(|t_1 - t_0|^2), \vec{\beta} \rangle}{(t_1 - t_0)^2} \right|}{\left| \frac{\vec{r}(t_1) - \vec{r}(t_0)}{t_1 - t_0} \right|^2}.$$

Since the limit of the denominator for $t_1 \to t_0$ is equal to $|\vec{r}'(t_0)|^2$ and since by the condition of the theorem it is nonzero, from the condition $\lim_{t_1 \to t_0} \frac{h}{d^2} = 0$ it follows firstly that $\langle \vec{r}'(t_0), \vec{\beta} \rangle = 0$, and then $\langle \vec{r}''(t_0), \vec{\beta} \rangle = 0$. To prove now the existence of an osculating plane, consider two cases:

$$(1) \quad \vec{r}'(t_0) \times \vec{r}''(t_0) \neq 0, \qquad (2) \quad \vec{r}'(t_0) \times \vec{r}''(t_0) = 0.$$

In the *first case* define a vector $\vec{\beta} = \frac{\vec{r}'(t_0) \times \vec{r}''(t_0)}{|\vec{r}'(t_0) \times \vec{r}''(t_0)|}$, and in the *second case* take for $\vec{\beta}$ an arbitrary unit vector orthogonal to $\vec{r}'(t_0)$. In *both cases* we have

$$\langle \vec{r}'(t_0), \vec{\beta} \rangle = \langle \vec{r}''(t_0), \vec{\beta} \rangle = 0.$$

Let α be the plane passing through the point $P = \vec{r}(t_0)$ and orthogonal to the vector $\vec{\beta}$. Then

$$h = |\langle \vec{o}(|t_1 - t_0|^2), \vec{\beta} \rangle|, \qquad d = |\vec{r}'(t_0)(t_1 - t_0) + \vec{o}(|t_1 - t_0|)|.$$

From this it follows that

$$\lim_{t_1 \to t_0} \frac{h}{d^2} = \lim_{t_1 \to t_0} \frac{\left| \frac{\langle \vec{o}(|t_1 - t_0|^2), \vec{\beta} \rangle}{(t_1 - t_0)^2} \right|}{\left| \frac{\vec{r}(t_1) - \vec{r}(t_0)}{t_1 - t_0} \right|^2} = \frac{\lim_{t_1 \to t_0} \left\langle \frac{\vec{o}(|t_1 - t_0|^2)}{|t_1 - t_0|^2}, \vec{\beta} \right\rangle}{|\vec{r}'(t_0)|^2} = 0.$$

Consequently, α is an osculating plane. Besides, as we see, in the *first case* the osculating plane is unique, and in the *second case* any plane containing a tangent line to γ at $P = \vec{r}(t_0)$ is an osculating plane. For a plane curve, the osculating plane is the plane containing φ. $\square$

We now deduce the equation of the osculating plane for the case that a curve is given by parametric equations and the vectors $\vec{r}'(t_0)$ and $\vec{r}''(t_0)$ at a given point $P = \vec{r}(t_0)$ are linearly independent. In this case the normal vector to the osculating plane, $\vec{\beta}$, as follows from Theorem 1.3.2, may be taken as $\vec{r}'(t_0) \times \vec{r}''(t_0)$,

$$\vec{\beta} = (\bar{y}'\bar{z}'' - \bar{y}''\bar{z}')(t_0)\vec{i} + (\bar{z}'\bar{x}'' - \bar{z}''\bar{x}')(t_0)\vec{j} + (\bar{x}'\bar{y}'' - \bar{x}''\bar{y}')(t_0)\vec{k},$$

and we obtain the equation of an osculating plane α:

$$A(x - x(t_0)) + B(y - y(t_0)) + C(z - z(t_0)) = 0,$$

where $A = y'z'' - y''z'$, $B = z'x'' - z''x'$, $C = x'y'' - x''y'$ are derived for $t = t_0$. Projecting γ orthogonally onto an osculating plane α, we obtain a plane curve $\bar{\gamma}$ of "minimal deviation" from γ. The value of this deviation has order slightly more than d^2. In detail, the lengths of the curves γ and $\bar{\gamma}$ that belong to the ball $B_P(d)$ (with center P and radius d) differ from each other by a value whose order is slightly greater than d^2.

At a point $P = \vec{r}(t)$ of a curve, where an osculating plane is unique, one may select among all normal directions a unique normal vector $\vec{v}$ by the conditions

(1) $\vec{v}$ is orthogonal to $\vec{r}'(t_0)$,

(2) $\vec{v}$ is parallel to an osculating plane,

(3) $\vec{v}$ forms an acute angle with the vector $\vec{r}''(t_0)$,

(4) $\vec{v}$ has unit length: $|\vec{v}| = 1$.

Such a vector $\vec{v}$ is the *principal normal vector* to a curve γ at a point P. It is easy to see that $\vec{v}$ can be expressed by the formula

$$\vec{v} = -\frac{\langle \vec{r}', \vec{r}'' \rangle}{|\vec{r}'| \cdot |\vec{r}' \times \vec{r}''|} \cdot \vec{r}' + \frac{|\vec{r}'|}{|\vec{r}' \times \vec{r}''|} \cdot \vec{r}''. \tag{1.8}$$

A principal normal vector $\vec{v}$ is defined invariantly in the sense that its direction does not depend on the choice of a curve γ parameterization. Let $\vec{r} = \vec{R}(\tau)$ be another parameterization of γ. Then, as we know, there is a function $t = t(\tau)$ such that $\vec{r}(t(\tau)) = \vec{R}(\tau)$ and

$$\vec{R}'_\tau = \vec{r}'_t \cdot t', \qquad \vec{R}''_{\tau\tau} = \vec{r}''_{tt} \cdot (t')^2 + \vec{r}'_t \cdot t''.$$

From these formulas it follows that $\langle \vec{v}, \vec{R}'_\tau \rangle = 0$ and $\langle \vec{v}, \vec{R}''_{\tau\tau} \rangle = \langle \vec{v}, \vec{r}''_{tt} \rangle \cdot (t')^2$, and consequently, the vector $\vec{v}$ satisfies all four conditions with respect to the parameterization $\vec{R}(\tau)$. Using the vectors $\vec{\tau} = \frac{\vec{r}'}{|\vec{r}'|}$ and $\vec{v}$, define a vector $\vec{\beta}$ by the formula $\vec{\beta} = \vec{\tau} \times \vec{\nu}$, and call it a *binormal vector*. The directions of $\vec{\tau}$ and $\vec{\beta}$ depend on the orientation of the curve and should be replaced by their opposites when the orientation is reversed. The vector $\vec{\nu}$, as was shown before, does not depend on the orientation of the curve.

In practice, it is convenient to derive $\vec{\tau}$, $\vec{\nu}$, and $\vec{\beta}$ in the following order: *first, the vector* $\vec{\tau} = \frac{\vec{r}'}{|\vec{r}'|}$, *then the vector* $\vec{\beta} = \frac{\vec{r}' \times \vec{r}''}{|\vec{r}' \times \vec{r}''|}$, *and finally the vector* $\vec{\nu} = \vec{\beta} \times \vec{\tau}$.

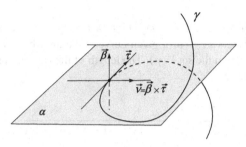

Figure 1.5. Tangent, principal normal, and binormal vectors.

1.4 Length of a Curve

Let γ be a closed arc of some curve, and $\vec{r} = \vec{r}(t)$ its parameterization; $a \le t \le b$. Note that a *polygonal line* is a curve in $\mathbb{R}^3$ ($\mathbb{R}^2$) composed of line segments passing through adjacent points of some ordered finite set of points $P_1, P_2, \ldots, P_k$. A polygonal line σ is a *regularly inscribed polygon in a curve* γ if there is a partition T of a line segment $[a, b]$ by the points $t_1 < t_2 < \cdots < t_k$ such that $\overrightarrow{OP_i} = \vec{r}(t_i)$. To each polygonal line there corresponds its *length* $l(\sigma)$ equal to $\sum_{i=1}^{k-1} P_i P_{i+1}$. Denote by $\Gamma(\gamma)$ the *set of all regularly inscribed polygonal lines in a curve* γ.

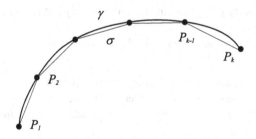

Figure 1.6. Regularly inscribed polygonal lines in a curve.

Definition 1.4.1. A continuous curve γ is called *rectifiable* if $\sup_{\sigma \in \Gamma(\gamma)} l(\sigma) < \infty$.

Definition 1.4.2. The *length* of a rectifiable curve γ is defined as the least upper bound of lengths of all regularly inscribed polygonal lines in a given curve γ: $l(\gamma) = \sup_{\sigma \in \Gamma(\gamma)} l(\sigma)$.

The next theorem gives us a sufficient condition for the existence of the length of a curve and the formula to calculate it.

Theorem 1.4.1. *A closed arc of any smooth curve is rectifiable, and its length is*

$$l(\gamma) = \int_a^b |\vec{r}'(t)|\, dt.$$

Proof. Let $\vec{r} = \vec{r}(t) = x(t)\vec{i} + y(t)\vec{j} + z(t)\vec{k}$ $(t \in [a, b])$ be a smooth parameterization of a closed arc γ of a given curve. Take an arbitrary polygonal line $\sigma: P_1, P_2, \ldots, P_k$ from the set $\Gamma(\gamma)$. The length of the ith segment of the polygonal line σ is equal to

$$P_i P_{i+1} = |\vec{r}(t_{i+1}) - \vec{r}(t_i)|$$
$$= \sqrt{[x(t_{i+1}) - x(t_i)]^2 + [y(t_{i+1}) - y(t_i)]^2 + [z(t_{i+1}) - z(t_i)]^2}.$$

Applying Lagrange's formula to each of the functions $x(t)$, $y(t)$ and $z(t)$, we obtain

$$P_i P_{i+1} = \sqrt{[x'(\xi_i)]^2 + [y'(\eta_i)]^2 + [z'(s_i)]^2}\, \Delta t_i, \tag{1.9}$$

where $t_i \leq \xi_i \leq t_{i+1}$, $t_i \leq \eta_i \leq t_{i+1}$, $t_i \leq s_i \leq t_{i+1}$, $\Delta t_i = t_{i+1} - t_i$. Since the functions $x'(t)$, $y'(t)$ and $z'(t)$ are continuous on a closed interval $[a, b]$, by Weierstrass's first theorem they are bounded on this closed interval; i.e., there is a real M such that $|x'(t)| < M$, $|y'(t)| < M$, and $|z'(t)| < M$, for all $t \in [a, b]$. Using the last inequality we obtain

$$l(\sigma) = \sum_{i=1}^{k-1} P_i P_{i+1} \leq \sqrt{3}M \sum_{i=1}^{k-1} \Delta t_i = \sqrt{3}M(b - a).$$

Since σ is an arbitrary polygonal line from the set $\Gamma(\gamma)$, it follows that

$$\sup_{\sigma \in \Gamma(\gamma)} l(\sigma) \leq \sqrt{3}M(b - a) < \infty.$$

The proof of the first statement of the theorem is complete.

Now we shall prove the second statement of the theorem.

To each polygonal line $\sigma: P_1, P_2, \ldots, P_k$ regularly inscribed in γ there corresponds some partition

$$T(\sigma): t_1 < t_2 < \cdots < t_k$$

of a closed interval $[a, b]$, and conversely, to each partition $T: t_1 < t_2 < \cdots < t_k$ of a closed interval $[a, b]$ there corresponds a polygonal line $\sigma(T): P_1, P_2, \ldots, P_k$, where P_i is the endpoint of the vector $\vec{r}(t_i)$. For each polygonal line $\sigma(t)$ define a number $\delta(T) = \max_{i=1 \ldots k-1} \Delta t_i$. We now prove that for any real $\varepsilon > 0$ there is a partition $T: t_1 < t_2 < \cdots < t_k$ of the line segment $[a, b]$ for which the following inequalities hold simultaneously:

$$|l(\gamma) - l(\sigma(T))| \leq \frac{\varepsilon}{3}, \tag{1.10}$$

$$\left| l(\sigma(T)) - \sum_{i=1}^{k-1} |\vec{r}'(t_i)| \Delta t_i \right| \leq \frac{\varepsilon}{3}, \tag{1.11}$$

$$\left| \sum_{i=1}^{k-1} |\vec{r}'(t_0)| \Delta t_i - \int_a^b |\vec{r}'(t)| dt \right| \leq \frac{\varepsilon}{3}. \tag{1.12}$$

Directly from the definition of the length of the curve γ and that it is rectifiable follows the existence of a partition T_1 of the line segment $[a, b]$ such that inequality (1.10) holds. The sum

$$\sum_{i=1}^{k-1} |\vec{r}'(t_i)| \Delta t_i$$

is a Riemann integral sum for the integral $\int_a^b |\vec{r}'(t)| dt$. Thus there is a real number $\delta_0 > 0$ such that for every partition T of a line segment $[a, b]$ with the property $\delta(T) < \delta_0$, the inequality (1.12) holds. Now take a partition T_2 of the line segment $[a, b]$ refining the partition T_1 and satisfying the inequality (1.12). For the partition T_2, in view of the triangle inequality, the inequalities (1.10) and (1.12) hold. The functions $x'(t)$, $y'(t)$, and $z'(t)$ are continuous and hence uniformly continuous on $[a, b]$. Thus for any real $\varepsilon_1 > 0$ there is a real number $\delta_1 > 0$ such that for $|t'' - t'| < \delta_1$, the inequalities

$$|x'(t'') - x'(t')| < \varepsilon_1, \quad |y'(t'') - y'(t')| < \varepsilon_1, \quad |z'(t'') - z'(t')| < \varepsilon_1$$

hold. Now take a partition T_3 of the line segment $[a, b]$ refining the partition T_2 and satisfying the inequality $\delta(T_3) \leq \min\{\delta_0, \delta_1\}$. For the ith segment $P_i P_{i+1}$ of such a partition we have

$$\left| P_i P_{i+1} - |\vec{r}'(t_i)| \cdot \Delta t_i \right|$$
$$= \left| \sqrt{[x'(\xi_i)]^2 + [y'(\eta_i)]^2 + [z'(\zeta_i)]^2} - \sqrt{[x'(t_i)]^2 + [y'(t_i)]^2 + [z'(t_i)]^2} \right| \Delta t_i$$
$$\leq \sqrt{[x'(\xi_i) - x'(t_i)]^2 + [y'(\eta_i) - y'(t_i)]^2 + [z'(\zeta_i) - z'(t_i)]^2} \Delta t_i \leq \sqrt{3} \varepsilon_1 \Delta t_i,$$

where the next-to-last inequality holds in view of the triangle inequality. Summing up these inequalities, we obtain

$$\left| l(\sigma(T_3)) - \sum_{i=1}^{n(T_3)-1} |\vec{r}'(t_i)| \Delta t_i \right| \leq \sqrt{3} \varepsilon_1 (b - a),$$

where $n(T_3)$ is the number of segments of the partition T_3. Select ε_1 satisfying the inequality $\sqrt{3} \varepsilon_1 (b - a) < \frac{\varepsilon}{3}$. Thus, if we take the partition T_3 in the role of the partition T of the line segment $[a, b]$, then the inequalities (1.10)–(1.12) will be satisfied simultaneously. Thus, summing up these inequalities, we obtain

$$\left| l(\gamma) - \int_a^b |\vec{r}'(t)| dt \right| \leq \frac{\varepsilon}{3} + \frac{\varepsilon}{3} + \frac{\varepsilon}{3} = \varepsilon. \tag{1.13}$$

Since $\varepsilon > 0$ is an arbitrary real number, the proof of second part of the theorem is complete. $\qquad\square$

If a curve γ is piecewise smooth, then its length can be calculated as a sum of lengths of its smooth parts. However, any piecewise regular curve has a smooth (nonregular!) parameterization (prove).

An arbitrary curve is called *rectifiable* if every one of its closed arcs is rectifiable. For rectifiable curves one can define the so-called *arc length parameterization*, which is based on the existence of the length of each closed arc. Let γ be an oriented rectifiable curve. Take an arbitrary point $P_0 \in \gamma$ and associate with P_0 the zero value of a parameter s. To any other point $P \in \gamma$ there corresponds the value of the parameter s that is equal to the arc length $P_0 P$ of the curve γ taken with the sign $(+)$ if P follows P_0, and with the sign $(-)$ if P precedes of P_0. If γ admits a smooth regular parameterization $\vec{r} = \vec{r}(t)$, then its arc length parameterization is also smooth and regular. Indeed, by taking into account the sign, we derive an arc length $P_0 P = s(t) = \int_0^t |\vec{r}(t)| \, dt$. The function $s(t)$ is differentiable and $\frac{ds}{dt} = |\vec{r}(t)| > 0$. Hence, there is an inverse function $t = t(s)$ and

$$\frac{dt}{ds} = \frac{1}{|\vec{r}'(t(s))|}. \tag{1.14}$$

The *arc length (or unit speed) parameterization of a curve* $\gamma : \vec{r} = \vec{r}(s)$ is defined by the formula

$$\vec{r}(s) = \vec{r}(t(s)). \tag{1.15}$$

From (1.15) follows the differentiability of the vector function $\vec{r}(s)$ and

$$|\vec{r}'(s)| = \left| \vec{r}'(t) \cdot \frac{dt}{ds} \right| = \frac{|\vec{r}'(t)|}{|\vec{r}'(t)|} = 1. \tag{1.16}$$

The last formula shows us that the given arc length parameterization is regular. For the arc length parameterization $\vec{r} = \vec{r}(s)$, the formulas for a tangent vector $\vec{\tau}$, a principal normal vector $\vec{\nu}$, and a binormal vector $\vec{\beta}$ take the simplest form:

$$\vec{\tau}(s) = \vec{r}'(s), \quad \vec{\nu}(s) = \frac{\vec{r}''(s)}{|\vec{r}''(s)|}, \quad \vec{\beta}(s) = \frac{\vec{r}'(s) \times \vec{r}''(s)}{|\vec{r}''(s)|}. \tag{1.17}$$

In fact, the first formula follows from (1.16) and the second from the equality

$$\langle \vec{r}'(s), \vec{r}'(s) \rangle' = 2\langle \vec{r}'(s), \vec{r}''(s) \rangle = 0.$$

From this it follows that $\vec{r}''(s)$ is orthogonal to the vector $\vec{r}'(s)$, and finally, the last formula follows from the definition of the vector $\vec{\beta}(s)$.

1.4.1 Formulas for Calculations

1. If $\gamma : \vec{r} = \vec{r}(t) = x(t)\vec{i} + y(t)\vec{j} + z(t)\vec{k}, \ a \leq t \leq b$, then

$$l(\gamma) = \int_a^b |\vec{r}'(t)| \, dt = \int_a^b \sqrt{x'^2 + y'^2 + z'^2} \, dt.$$

2. If $\gamma: y = f_1(x), z = f_2(x), \ a \le x \le b$, then

$$l(\gamma) = \int_a^b \sqrt{1 + f_1'^2 + f_2'^2} \, dx.$$

3. If $\gamma(t): \vec{r} = \vec{r}(t) = x(t)\vec{i} + y(t)\vec{j}$, a plane curve, then

$$l(\gamma) = \int_a^b \sqrt{x'^2 + y'^2} \, dt. \tag{1.18}$$

4. If $\gamma: y = f(x), \ a \le x \le b$, then

$$l(\gamma) = \int_a^b \sqrt{1 + f'^2} \, dx.$$

Example 1.4.1. (a) Consider a *helix* $\vec{r} = [a\cos t, a\sin t, bt]$ with $\vec{r}' = [-a\sin t, a\cos t, b]$, which in view of $\angle(\vec{r}'(t), OZ) = $ const is also called a *curve of a constant slope*. The speed is $|\vec{r}'(t)| = \sqrt{a^2 + b^2} = c$. Then $s(t) = \int_0^t |\vec{r}'(t)| \, dt = ct$. So an arc length parameterization is given by $\vec{r}_1(s) = [a\cos\frac{s}{c}, a\sin\frac{s}{c}, b\frac{s}{c}]$. Finally, we compute the arc length of the helix period $L = \int_0^{2\pi} |\vec{r}'(t)| \, dt = 2\pi\sqrt{a^2 + b^2}$. The length of the circle $L = 2\pi a$ is the particular case $b = 0$.

(b) For a parabola $\vec{r} = [t, t^2/2]$ (a very simple curve geometrically) we obtain $s(t) = \int_0^t \sqrt{1 + t^2} \, dt = (t\sqrt{1 + t^2} + \ln(t + \sqrt{1 + t^2}))/2$. However, it is a difficult task to find $t = t(s)$ from this equation.

1.5 Problems: Convex Plane Curves

We review some notions from the theory of convex plane curves. Recall that a closed region $D \subset \mathbb{R}^2$ is *convex* if for every pair of its points A and B it contains the entire line segment AB joining these points: $A \in D, B \in D \Rightarrow AB \subset D$. A connected boundary component of a convex region is called a *convex curve*. Another definition of a convex curve that is equivalent to above given can be formulated as follows: a curve γ is *convex* if each of its points has a support line. A straight line a through a point P of a curve γ is a *support line* to γ at $P \in \gamma$ if the curve is located entirely in one of the two half-planes determined by a. A tangent line need not exist at each point of a convex curve, but for the points, where the tangent line exists, it is also a support line.

Now we shall formulate and solve some problems about convex curves.

Problem 1.5.1. Every closed convex curve has length (i.e., it is a rectifiable curve).

Solution. Let $\sigma: P_1, P_2, \ldots, P_k = P_1$ be an arbitrary closed polygonal line regularly inscribed in a convex curve γ. If we pass a support line to γ through a point P_i, then the points P_{i-1} and P_{i+1} are located on one side of this straight line, and hence the inner angle of a polygonal line σ at a vertex P_i is not greater than π. Consequently, a polygonal line σ is convex. Since γ is a closed curve, then there is a triangle Δ containing it, and hence containing σ, and from this follows the inequality $l(\sigma) \leq l(\Delta)$. Since σ is an arbitrary regularly inscribed polygonal line in γ, we have $l(\gamma) = \sup_{\sigma \in \Gamma(\gamma)} l(\sigma) \leq l(\Delta)$. □

It turns out that the length of a closed convex curve can be calculated using its orthogonal projections onto all straight lines through an arbitrary fixed point. Denote by $a(\varphi)$ the straight line through the coordinate system's origin forming an angle φ with the OX axis, and by $d_\gamma(\varphi)$ the length of the orthogonal projection of the curve γ onto the straight line $a(\varphi)$.

Problem 1.5.2. Deduce the formula $l(\gamma) = \int_0^\pi d_\gamma(\varphi)\, d\varphi$.

Solution. Take an arbitrary line segment μ of length d. Without loss of generality one may assume that it is located on the OY axis. Then $d_\mu(\varphi) = d \sin\varphi$ and

$$\int_0^\pi d_\mu(\varphi)\, d\varphi = \int_0^\pi d \sin\varphi\, d\varphi = d(-\cos\varphi)|_0^\pi = 2d.$$

Now let $\sigma: P_1, P_2, \ldots, P_k = P_1$ be an arbitrary convex closed polygonal line. Then $d_\sigma(\varphi) = \frac{1}{2}\sum_{i=1}^{k-1} d_{P_i P_{i+1}}(\varphi)$ and

$$\int_0^\pi d_\sigma(\varphi)\, d\varphi = \frac{1}{2}\sum_{i=1}^{k-1}\int_0^\pi d_{P_i P_{i+1}}(\varphi)\, d\varphi = \sum_{i=1}^{k-1} P_i P_{i+1} = l(\sigma).$$

By the way, we have proved our formula for polygonal lines. For an arbitrary convex curve the formula of the problem follows from the previous formula and from the definition of the length of a curve. □

Problem 1.5.3. Let $\gamma_1: \vec{r}_1 = \vec{r}_1(s)$ and $\gamma_2: \vec{r}_2 = \vec{r}_2(s)$ be smooth curves in $\mathbb{R}^3$, s the arc length parameter. Denote by $l(s)$ the length of a segment $\gamma_1(s)\gamma_2(s)$. Then we have the formula

$$\frac{dl}{ds} = \cos\alpha_1(s) + \cos\alpha_2(s),$$

where $\alpha_1(s)$ and $\alpha_2(s)$ are the angles between the vector $\overrightarrow{\gamma_2(s)\gamma_1(s)}$ and the vectors $\vec{\tau}_1 = \frac{d\vec{r}_1}{ds}$, $\overrightarrow{\gamma_1(s)\gamma_2(s)}$, and $\vec{\tau}_2 = \frac{d\vec{r}_2}{ds}$, respectively.

Solution. If the equations of the curves γ_1 and γ_2 are written in the parametric form $x_1 = x_1(s)$, $y_1 = y_1(s)$, $z_1 = z_1(s)$ and $x_2 = x_2(s)$, $y_2 = y_2(s)$, $z_2 = z_2(s)$, respectively, then

$$l(s) = \sqrt{(x_2 - x_1)^2 + (y_2 - y_1)^2 + (z_2 - z_1)^2}$$

and

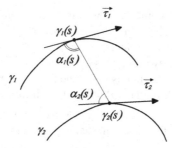

Figure 1.7. First variation of the distance $l(s) = |\gamma_1(s)\gamma_2(s)|$.

$$\frac{dl}{ds} = \frac{(x_2 - x_1)(x_2' - x_1') + (y_2 - y_1)(y_2' - y_1') + (z_2 - z_1)(z_2' - z_1')}{\sqrt{(x_2 - x_1)^2 + (y_2 - y_1)^2 + (z_2 - z_1)^2}}$$

$$= \frac{(x_1 - x_2)x_1' + (y_1 - y_2)y_1' + (z_1 - z_2)z_1'}{l(s)}$$

$$+ \frac{(x_2 - x_1)x_2' + (y_2 - y_1)y_2' + (z_2 - z_1)z_2'}{l(s)}$$

$$= \left\langle \frac{\overrightarrow{\gamma_2\gamma_1}}{|\gamma_2\gamma_1|}, \vec{\tau}_1 \right\rangle + \left\langle \frac{\overrightarrow{\gamma_1\gamma_2}}{|\gamma_1\gamma_2|}, \vec{\tau}_2 \right\rangle = \cos\alpha_1(s) + \cos\alpha_2(s).$$

In the particular case that γ_2 degenerates to a point, we have $\frac{dl}{ds} = \cos\alpha_1(s)$. If the curves γ_1 and γ_2 are parameterized by an arbitrary parameter t, and $l(t) = \gamma_1(t)\gamma_2(t)$, then

$$\frac{dl}{dt} = \cos\alpha_1(t)\frac{ds_1}{dt} + \cos\alpha_2(t)\frac{ds_2}{dt},$$

where $s_1(t) = \int_0^t |\vec{r}_1(t)|\,dt$ and $s_2(t) = \int_0^t |\vec{r}_2(t)|\,dt$. $\qquad\square$

Problem 1.5.4. Let γ be an arc of a smooth convex curve with endpoints A_1 and A_2. Denote by $l(h)$ the length of a chord $A_1(h)A_2(h)$ on γ that is parallel to the straight line A_1A_2 and is located at distance h from it. Denote by $\alpha_1(h)$ and $\alpha_2(h)$ the angles that the chord $A_1(h)A_2(h)$ forms with γ. Then the following formula holds:

$$\frac{dl}{dh} = \cot\alpha_1(h) + \cot\alpha_2(h).$$

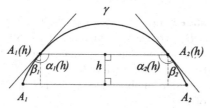

Figure 1.8. Derivative of the length of a chord $A_1(h)A_2(h)$.

Solution. Denote by B a point on γ where the tangent line is parallel to the straight line $A_1 A_2$. The point B divides γ onto two arcs: γ_1 from A_1 to B and γ_2 from A_2 to B. Let $\vec{r}_1 = \vec{r}_1(s)$ and $\vec{r}_2 = \vec{r}_2(s)$ be the arc length parameterizations of these arcs. Define two functions $h_1(s)$ and $h_2(s)$ on the curves γ_1 and γ_2 equal to the distances from the points $\vec{r}_1(s)$ and $\vec{r}_2(s)$, respectively, to the straight line $A_1 A_2$. Then from the formula of Problem 1.5.3 it follows that $\frac{dh_1}{ds} = \cos \beta_1(s)$ and $\frac{dh_2}{ds} = \cos \beta_2(s)$, where $\beta_1 = \alpha_1 - \frac{\pi}{2}$ and $\beta_2 = \alpha_2 - \frac{\pi}{2}$ or $\frac{dh_1}{ds} = \sin \alpha_1(s)$, $\frac{dh_2}{ds} = \sin \alpha_2(s)$. From the formula of the same problem it follows that

$$\frac{dl}{dh} = \cos \alpha_1 \frac{ds}{dh_1} + \cos \alpha_2 \frac{ds}{dh_2} = \frac{\cos \alpha_1}{\sin \alpha_1} + \frac{\cos \alpha_2}{\sin \alpha_2} = \cot \alpha_1 + \cot \alpha_2. \qquad \square$$

In the rest of this section we will discuss the isoperimetric problem.

Problem 1.5.5 (Isoperimetric problem). Among all closed curves with fixed length find one bounding a region of a maximal area.

This statement can be also reformulated in the following form: Let l be the length of some closed curve γ, and let S be the area of a region $D(\gamma)$ bounded by γ. Then any closed curve satisfies the *isoperimetric inequality* $S \leq l^2/4\pi$, and equality holds if and only if γ is a circle.

Solution. Solve this problem under the assumption that an extremal curve exists. Let γ be an extremal curve; i.e., a curve with length l, bounding a region of a maximal area. Then it has the following properties:

(1) γ is a convex curve;
(2) if the points A_1 and A_2 divide γ into two arcs of equal lengths, then the chord $A_1 A_2$ divides $D(\gamma)$ into two regions D_1 and D_2 of equal areas.

Proof of the first statement. Assume that the curve γ is not convex. Then there exist two different points B_1 and B_2 on γ such that γ is entirely located on one side of the straight line $B_1 B_2$, and the interior points of the line segment $B_1 B_2$ do not belong to γ. The points B_1 and B_2 divide γ into two arcs γ_1 and γ_2. Together with the line segment $B_1 B_2$ they form two closed curves σ_1 and σ_2, one of which, say σ_1, belongs to the region bounded by the other curve, σ_2. Take a curve $\overline{\gamma}_1$ that is symmetric to γ_1 with respect to the straight line $B_1 B_2$. Then $\overline{\gamma} = \overline{\gamma}_1 \cup \gamma_2$ is a closed curve with the same length l that bounds a region $D(\overline{\gamma}) \supset D(\gamma)$, and $S(D(\overline{\gamma})) > S(D(\gamma))$ holds, which contradicts the extremality of γ.

Proof of the second statement. Assume that the line segment $A_1 A_2$ divides $D(\gamma)$ into two regions D_1 and D_2 with unequal areas. Suppose that $S(D_2) > S(D_1)$. Denote by $\overline{D}_2$ the region that is symmetric to D_2 with respect to the straight line $A_1 A_2$. The area of the region $D = D_2 + \overline{D}_2$ is greater than the area of $D(\gamma)$, and the length of its boundary curve is l, which again contradicts the extremal property of γ.

We finish the solution of the problem with an elegant and beautiful argument by E. Steinitz. Let A_1 and A_2 be two points dividing γ into two arcs with equal

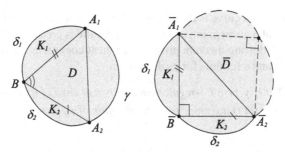

Figure 1.9. Isoperimetric problem solution.

lengths. Take an arbitrary point $B \neq A_1$, $B \neq A_2$. Then $\angle A_1 B A_2 = 90°$. Actually, assume that there exists a point $B \in \gamma$ such that $\angle A_1 B A_2 \neq 90°$. Denote by δ_1 the arc $A_1 B_0$ of γ, and by δ_2 the arc $B A_2$. The arc δ_1 and the chord $A_1 B$ bound a region K_1, and the arc δ_2 together with the chord $B A_2$ bound a region K_2.

Now construct a triangle $\triangle \bar{A}_1 \bar{B} \bar{A}_2$ for which $\bar{A}_1 \bar{B} = A_1 B$, $\bar{A}_2 \bar{B} = A_2 B$, and $\angle \bar{A}_1 \bar{B} \bar{A}_2 = 90°$. Then the inequality $S(\triangle \bar{A}_1 \bar{B} \bar{A}_2) > S(\triangle A_1 B A_2)$ holds. Now construct a region $\overline{D}_1$ placing the regions K_1, K_2, respectively, along the legs of $\triangle \bar{A}_1 \bar{B} \bar{A}_2$, and a region $\overline{D}_2$ that is symmetric to it with respect to the straight line $A_1 A_2$. The boundary of the region $\overline{D} = \overline{D}_1 \cup \overline{D}_2$ has the same length l, but its area is greater than the area of $D(\gamma)$, which is a contradiction. Consequently, for any point $B \in \gamma$ with the property $B \neq A_1$, $B \neq A_2$ we have $\angle A_1 B A_2 = 90°$, but from this it follows that γ is a circle. The radius R of this circle is $l/2\pi$, and its area is $S(D(\gamma)) = l^2/4\pi$; i.e., $S = l^2/4\pi$.   □

1.6 Curvature of a Curve

Let γ be a smooth curve in $\mathbb{R}^3$. Take on it a point P and another point P_1. Denote by Δs the arc length $P P_1$ on γ and by $\Delta \theta$ the angle between tangent vectors $\vec{\tau}$ and $\vec{\tau}_1$ to γ at P and P_1.

Definition 1.6.1. The limit

$$\lim_{P_1 \to P} \frac{\Delta \theta}{\Delta s} = \lim_{\Delta s \to 0} \frac{\Delta \theta}{\Delta s},$$

if it exists, is called the *curvature* of the curve γ at the point P.

The curvature of a curve γ at a point $\gamma(t)$ will be denoted by $k(t)$.

Example 1.6.1. (a) Let γ be a straight line. Then $\Delta \theta \equiv 0$, and $k = 0$ holds at all points of γ. (b) Let γ be a circle of radius R. Then $\Delta s = R \Delta \theta$ and $\lim_{\Delta s \to 0} \frac{\Delta \theta}{\Delta s} = \frac{1}{R}$; i.e., the curvature of a circle is the same number $\frac{1}{R}$ at any of its points.

Later on, we shall prove that there are no plane curves of constant curvature other than circles and straight lines. Definition 1.6.1 and Example 1.6.1 show us that the curvature of a curve is the measure of its deviation from a straight line in a neighborhood of a given point, and that the curvature is greater as this deviation is greater. The following theorem gives us sufficient conditions for the existence of the curvature and a formula for its derivation.

Theorem 1.6.1. *Let γ be a C^2-regular curve. Then at each of its points there is a curvature. If $\vec{r} = \vec{r}(t)$ is a regular parameterization of γ, then $k = \frac{|\vec{r}' \times \vec{r}''|}{|\vec{r}'|^3}$.*

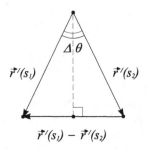

$$\vec{r}'(s_1) - \vec{r}'(s_2)$$

Figure 1.10. Triangle composed from the vectors $\vec{r}'(s_1)$ and $\vec{r}'(s_2)$.

Proof. Let $\vec{r} = \vec{r}(s)$ be the arc length parameterization of γ, and let $P_1 = \vec{r}(s_1)$, $P_2 = \vec{r}(s_2)$. Then $\Delta s = |s_2 - s_1|$, and $\Delta\theta$ is the angle between vectors $\vec{r}'(s_1)$ and $\vec{r}'(s_2)$. Since $|\vec{r}'(s_1)| = |\vec{r}'(s_2)| = 1$, then $2\sin\frac{\Delta\theta}{2} = |\vec{r}'(s_1) - \vec{r}'(s_2)|$. Thus

$$\lim_{s\to 0}\frac{\Delta\theta}{\Delta s} = \lim_{\Delta\theta\to 0}\frac{\Delta\theta}{2\sin\frac{\Delta\theta}{2}} \cdot \lim_{s\to 0}\frac{|\vec{r}'(s_1) - \vec{r}'(s_2)|}{\Delta s} = |\vec{r}''(s_1)|.$$

By these arguments, the first part of the theorem has been proved. Moreover, we have the formula

$$k = |\vec{r}''(s)| \tag{1.19}$$

at a point $\gamma(s)$. Now let $\vec{r} = \vec{r}(t)$ be an arbitrary regular parameterization of γ. Then

$$\vec{r}'_s = \vec{r}'_t\frac{dt}{ds} = \vec{r}'_t(t)\frac{1}{|\vec{r}'(t)|},$$

$$\vec{r}''_{ss} = \vec{r}''_{tt}\left(\frac{dt}{ds}\right)^2 + \vec{r}'_t\frac{d^2t}{ds^2} = \frac{\vec{r}''_{tt}}{|\vec{r}'_t|^2} - \vec{r}'_t\frac{\langle\vec{r}''_{tt}, \vec{r}'_t\rangle}{|\vec{r}'_t|^4},$$

and

$$k^2 = |\vec{r}''_{ss}|^2 = \frac{\langle\vec{r}''_{tt}, \vec{r}''_{tt}\rangle}{|\vec{r}'|^4} - \frac{2\langle\vec{r}''_{tt}, \vec{r}'_t\rangle^2}{|\vec{r}'|^6} + \frac{\langle\vec{r}''_{tt}, \vec{r}'_t\rangle^2}{|\vec{r}'|^6}$$

$$= \frac{|\vec{r}''_{tt}|^2 \cdot |\vec{r}'_t|^2 - \langle\vec{r}''_{tt}, \vec{r}'_t\rangle^2}{|\vec{r}'|^6} = \frac{|\vec{r}'_t \times \vec{r}''_{tt}|^2}{|\vec{r}'_t|^6}.$$

From this follows

$$k = \frac{|\vec{r}_{tt}'' \times \vec{r}_t'|}{|\vec{r}_t'|^3}.$$

□

The last formula shows us that the noncollinearity condition of the vectors $\vec{r}_t'$ and $\vec{r}_{tt}''$ has a geometrical sense; i.e., it does not depend on the choice of a parameterization. If at some point of γ the curvature is nonzero, then $\vec{r}_t'$ and $\vec{r}_{tt}''$ are nonparallel and conversely.

This remark allows us to give a geometric condition for the uniqueness of the osculating plane to a curve γ at any of its points P, and to complete Theorem 1.3.2 by the following statement.

Theorem 1.6.2. *A necessary and sufficient condition for the existence of a unique osculating plane to a C^2-regular curve γ at any of its points is that the curvature of γ be nonzero at this point.*

As we have just shown, a straight line has zero curvature at each of its points. The converse statement is also true: if the curvature of a curve γ at each of its point is zero, then γ is a straight line. Indeed, if $k \equiv 0$, then $\vec{r}_{ss}'' \equiv 0$, from which follows $\vec{r}_s' = \vec{c}_1$ and $\vec{r} = \vec{c}_1 s + \vec{c}_2$.

1.6.1 Formulas for Calculations

(1) If $\gamma : \vec{r} = \vec{r}(t) = x(t)\vec{i} + y(t)\vec{j} + z(t)\vec{k}$, then

$$k = \frac{\sqrt{(y'z'' - z'y'')^2 + (z'x'' - x'z'')^2 + (x'y'' - y'x'')^2}}{(x'^2 + y'^2 + z'^2)^{\frac{3}{2}}},$$

(2) If γ is a plane curve $\vec{r} = \vec{r}(t) = x(t)\vec{i} + y(t)\vec{j}$, then

$$k = \frac{|y''x' - x''y'|}{(x'^2 + y'^2)^{\frac{3}{2}}},$$

(3) If γ is a graph $y = f(x)$, then

$$k = \frac{|f''(x)|}{(1 + x'^2)^{\frac{3}{2}}}.$$

1.6.2 Plane Curves

The curvature of plane curves can be provided with a sign in the following way. Draw an arbitrary continuous normal vector field $\vec{n}(t)$ along a curve γ. Then the curvature of γ at a point $P = \vec{r}(t)$ is *positive* if the principal normal vector $\vec{v}(t)$ of γ coincides with $\vec{n}(t)$, and *negative* in the opposite case. For a closed simple curve γ, the normal vector field $\vec{n}(t)$ will be directed inside the region bounded

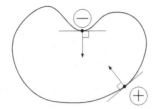

Figure 1.11. The plane curve sign of curvature.

by γ. In this case, the curvature of a curve γ is positive at a point if the region is convex "outwards," and the curvature is negative if it is convex "inwards."

In particular, by this definition of a curvature's sign, a closed convex curve has nonnegative curvature at each point. For oriented curves denote by $\alpha(t)$ the *angle between the vector $\vec{r}'(t)$ and the direction of the OX axis*, and define the *curvature's sign* as the rate at which the angle $\alpha(t)$ is changing, where we assume $\mathrm{sign}k = \mathrm{sign}\frac{d\alpha}{dt}$. The value of the angle $\alpha(t)$ is defined by the following method: the angle $\alpha(0)$ is equal to the angle measured from $\vec{i}$ counterclockwise with sign $(+)$, or clockwise with sign $(-)$. For other values of t the angle $\alpha(t)$ is defined by continuity; it increases when the vector $\vec{r}'(t)$ turns counterclockwise, and decreases otherwise.

In particular, if a plane curve γ is given explicitly as $y = f(x)$, then it naturally obtains an orientation (by increasing x-variable), and then the curvature's sign coincides with the sign of $f''(x)$; i.e., in this case, $k = \frac{f''}{(1+f'^2)^{3/2}}$. If a curve is given by the arc length parameterization, then $k(s) = \frac{d\alpha}{ds}$. If the curvature k of a curve at any of its points is nonzero, then the real number $1/|k|$ is called the *radius of curvature* of the curve at the given point and is denoted by R: $R = 1/|k|$. We say that the radius of curvature is infinite if the curvature is zero; moreover, the radius of curvature R can be is considered with the sign in correspondence with the formula $R = 1/k$.

Plane curves are uniquely determined by their curvature $k(s)$, given as a function of the arc length parameter s. But before we formulate this theorem, we shall generalize the definition of a curve given above.

Due to Definition 1.2.1, we represent a regular curve as the differentiable image of an open interval or a circle into $\mathbb{R}^2$ (or $\mathbb{R}^3$). Such a definition was sufficient for studying the local properties of a curve. However, when one studies the properties of a curve *as a whole*, inevitably there appear curves with points of self-intersection. Moreover, curves defined by their geometrical properties also often have points of self-intersection, for example, elongated cycloids and hypocycloids, lemniscates of Bernoulli, etc.

Thus, we shall hereinafter define a *curve* as a *locally diffeomorphic* image of an open interval I or a circle S^1 into $\mathbb{R}^2$ (or $\mathbb{R}^3$).

More precisely, two local diffeomorphisms $\varphi_1(t)$ and $\varphi_2(t)$ of an open interval or a circle into $\mathbb{R}^2$ (or $\mathbb{R}^3$) are *equivalent* if there is a diffeomorphism $t = t(\tau)$ of an open interval or a circle onto itself such that $\varphi_1(t(\tau)) \equiv \varphi_2(t)$.

The equivalency class of local diffeomorphisms of I or S^1 will be called a *smooth regular curve*. We shall treat a point of self-intersection of a smooth regular curve as two different points having two corresponding tangent vectors, two main normal vectors, two values of the curvature, etc. If a curve γ has no points of self-intersection, then it is *simple*.

Theorem 1.6.3. *Let $h(s)$ be an arbitrary continuous function on a line segment $[a, b]$. Then there is a unique (up to a rigid motion) curve γ for which $h(s)$ is the curvature function and s is the arc length parameter.*

Proof. Let the functions $x(s)$, $y(s)$, and $\alpha(s)$ satisfy the system of equations

$$\frac{dx}{ds} = \cos \alpha(s), \qquad \frac{dy}{ds} = \sin \alpha(s), \qquad \frac{d\alpha}{ds} = h(s).$$

Solving this system, we get

$$\alpha(s) = \alpha_0 + \int_0^s h(s)\,ds, \ \ x(s) = x_0 + \int_0^s \cos \alpha(s)\,ds, \ \ y(s) = y_0 + \int_0^s \sin \alpha(s)\,ds.$$

The obtained curve $\gamma: x = x(s)$, $y = y(s)$ satisfies all the conditions of the theorem. Prove that s is the arc length parameter. By formula (1.18) we have

$$l = \int_a^s \sqrt{(x')^2 + (y')^2}\,ds = \int_a^s ds = s - a.$$

Further, by formula (1.19),

$$|k(s)| = |x''(s)\vec{i} + y''(s)\vec{j}| = \sqrt{(x'')^2 + (y'')^2} = \sqrt{|\alpha'|^2} = \left|\frac{d\alpha}{ds}\right| = |h(s)|.$$

In view of the definition of the sign of curvature we obtain $k(s) = \frac{d\alpha}{ds} = h(s)$. Finally, the coordinates of the initial point on the curve $\gamma(s)$ are actually (x_0, y_0), and the direction of the tangent vector $\vec{\tau}(0)$ forms the angle α_0 with the OX axis. Hence, if there exist two curves with equal curvatures, then a rigid motion that matches their initial points and initial tangent vectors at this point also maps one curve to the other. $\qquad\square$

From Theorem 1.6.3 it immediately follows that if the curvature of a curve is constant, then the curve is either a line segment or an arc of a circle. The equation $k = k(s)$ is called a *natural equation* of a curve. A simple analysis of the proof of Theorem 1.6.3 shows that its statement remains true if a function $h(s)$ is only integrable. In particular, Theorem 1.6.3 holds if $h(s)$ is a piecewise continuous function with a finite number of discontinuity points of the first order. In this case, γ with the above-mentioned function of curvature would be a smooth regular curve having a finite number of arcs of class C^2.

To a point at which two arcs of class C^2 meet, correspond two values k_- and k_+ that are the left- and right-hand limits of the curvature function. We say that the curvature of a curve γ at this point is not smaller than k_0 (not greater than k_0) if $\min(k_-, k_+) \geq k_0$ ($\max(k_-, k_+) \leq k_0$).

In Section 1.7 if the opposite is not supposed, we shall refer to a curve γ from this class; i.e., a smooth regular curve with a piecewise continuous curvature function $k(s)$.

1.7 Problems: Curvature of Plane Curves

Let two plane curves γ_1 and γ_2 touch each other at a common point M, and let the curvature sign of γ_1 and γ_2 be defined using the same normal vector $\vec{n}$. Denote by k_1 and k_2 the curvatures of the curves γ_1 and γ_2 at M.

Problem 1.7.1. If $k_1 > k_2$, then there is a neighborhood U of a point M in which a curve γ_1 with the exception of the point M is located on one side of γ_2 defined by the direction of $\vec{n}$.

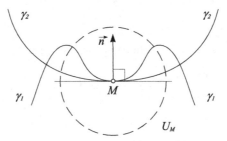

Figure 1.12. Curves with curvatures $k_1 > k_2$.

Solution. Introduce a rectangular coordinate system (x, y) on $\mathbb{R}^2$, locating its origin at M, and such that the direction of the OY axis coincides with $\vec{n}$. Then in some neighborhood V of a point M the equations of γ_1 and γ_2 can be reduced to the explicit form $y = f_1(x)$ and $y = f_2(x)$. From the conditions of the problem and the selection of a coordinate system it follows that $f_1(0) = f_2(0)$, $f_1'(0) = f_2'(0)$ and $k_1(0) = f_1''(0)$, $k_2(0) = f_2''(0)$. We apply Taylor's formula to the function $f(x) = f_1(x) - f_2(x)$. Since $f(0) = f'(0) = 0$,

$$f(x) = \frac{1}{2}x^2\left[(k_1 - k_2) + \frac{\bar{o}(x^2)}{x^2}\right].$$

From the last formula follows the existence of a real $\varepsilon > 0$ such that for $|x| < \varepsilon$, the expression $(k_1 - k_2) + \frac{\bar{o}(x^2)}{x^2}$ is positive.   □

If a curve γ is closed, then it divides the plane into two parts, one of which is compact. Denote this compact region by $D(\gamma)$.

Problem 1.7.2. Let $\gamma(s)\colon \vec{r} = \vec{r}(s)$ be an arc length parameterized plane curve of class C^2, and $\ell(s) = |\vec{r}(s)|$. Prove the formula

$$\frac{d^2\ell}{ds^2}\bigg|_{s=0} = k\cos\alpha + \frac{\cos^2\alpha}{\ell(0)},$$

where $k = k(0)$ is the curvature of the curve at $\gamma(0)$, and α the angle between the vector $\vec{r}_0 = \frac{\vec{r}(0)}{|\vec{r}(0)|}$ and the main normal $\vec{v} = \vec{v}(0)$.

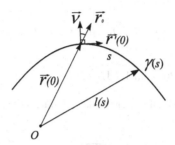

Figure 1.13. The second variation of $l(s) = |\vec{r}(s)|$.

Solution. Since $\ell = \ell(s) = \langle \vec{r}, \vec{r} \rangle^{\frac{1}{2}}$, then $\frac{d\ell}{ds} = \frac{\langle \vec{r}, \vec{r}' \rangle}{|\vec{r}|} = \langle \vec{r}_0, \vec{r}' \rangle$. From this we obtain

$$\frac{d^2\ell}{ds^2} = \frac{d}{ds}\left(\frac{\langle \vec{r}, \vec{r}' \rangle}{|\vec{r}|}\right) = \frac{\langle \vec{r}', \vec{r}' \rangle}{|\vec{r}|} + \frac{\langle \vec{r}, \vec{r}'' \rangle}{|\vec{r}|} - \frac{\langle \vec{r}, \vec{r}' \rangle^2}{|\vec{r}|^3}$$

$$= \frac{1}{\ell(0)}(1 - \langle \vec{r}_0, \vec{r}' \rangle^2) + k\langle \vec{r}_0, \vec{v} \rangle = k\cos\alpha + \frac{\cos^2\alpha}{\ell(0)}.$$

Note that the second variation formula is simplified if the origin of the coordinate system is placed so that the vectors $\vec{r}_0$ and $\vec{v}$ coincide. Then $\cos\alpha = 1$ and $\frac{d^2\ell}{ds^2}\big|_{s=0} = k + \frac{1}{\ell(0)}$. ☐

Problem 1.7.3 (Frenet formulas for plane curves). Prove that the formulas

$$\frac{dx}{ds} = \cos\alpha(s), \qquad \frac{dy}{ds} = \sin\alpha(s), \qquad \frac{d\alpha}{ds} = k$$

are equivalent to the equalities

$$\frac{d\vec{\tau}}{ds} = k(s)\vec{v}, \qquad \frac{d\vec{v}}{ds} = -k(s)\vec{\tau}.$$

Solution. Since $\vec{\tau} = \vec{\tau}(s) = \cos\alpha\,\vec{i} + \sin\alpha\,\vec{j}$, but $\vec{v} = \vec{v}(s) = -\sin\alpha\,\vec{i} + \cos\alpha\,\vec{j}$, then

$$\frac{d\vec{\tau}}{ds} = \alpha'(-\sin\alpha)\vec{i} + \alpha'\cos\alpha\,\vec{j} = \alpha'\vec{v} = k(s)\vec{v},$$

$$\frac{d\vec{v}}{ds} = -\alpha'\cos\alpha\,\vec{i} - \alpha'\sin\alpha\,\vec{j} = -\alpha'\vec{\tau} = -k(s)\vec{\tau}. \qquad ☐$$

Problem 1.7.4. If a curve γ is closed, then there is a point on it where the curvature is positive.

Solution. Let P be an arbitrary point in a region $D(\gamma)$. Take a sufficiently large real R such that a disk with center at P and radius R contains γ. Decrease the radius of this disk until the circle with center P and radius R_0 is for the first time tangent to γ at some point P_1. The curvature of the circle is $1/R_0$, but at this point, as follows from Problem 1.7.1, the curvature $k(P_1)$ of the curve γ is not smaller than $1/R_0$.   □

Problem 1.7.5. The curvature of a closed convex curve is nonnegative at each of its points.

Solution. Follows from Problem 1.7.1.   □

Problem 1.7.6. If a simple closed curve has a nonnegative curvature at each of its points, then it is convex.

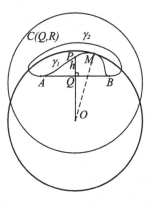

Figure 1.14. The curvature of a plane curve, Problem 1.7.6.

Solution. Assume that γ is not convex. Then there exist two points A and B on γ such that the line segment AB lies outside of $D(\gamma)$, and γ is located on one side of the straight line AB. The points A and B divide γ onto two arcs, γ_1 and γ_2. One of the curves $\sigma_1 = \gamma_1 \cup AB$ and $\sigma_2 = \gamma_2 \cup AB$ contains $D(\gamma)$. Assume that this curve is σ_2. Find a point P on the arc γ_1 with the maximal distance from the straight line AB. Denote by Q the top of the perpendicular dropped from P onto the straight line AB and let $h = PQ$, $b = \max(QA, QB)$. Let $C(Q, R)$ be the disk with center Q and radius R such that the inequality

$$R > \frac{h^2 + b^2}{2h}, \tag{1.20}$$

holds and is sufficiently large that $C(Q, R)$ contains σ_1. Now move the center O of this disk along the straight line PQ in the direction of the vector

$\overrightarrow{PQ}$ until $C(Q, R)$ touches the curve σ_1 at some point M. We shall prove that $M \in \gamma_1 \setminus \{A \cup B\}$. In fact, if $M = A$ or $M = B$, then $OQ = OP - PQ < R - h$, and consequently, $OQ^2 + b^2 < (R - h)^2 + b^2 < R^2$ in view of the inequality (1.20). But $OQ^2 + b^2 = R^2$, which is a contradiction. Hence $M \in \gamma \setminus \{A \cup B\}$. The curvature k_1 of γ_1 relative to σ_1 at M, in view of Problem 1.7.1, is not smaller than $1/R$, but with respect to γ it is equal to $-R_1 < -1/k$, contrary to the condition. $\qquad\square$

Problem 1.7.7. If γ is a simple closed curve, then

$$\int_\gamma k(s)\, ds = 2\pi.$$

Solution. Inscribe in a curve γ a closed polygonal line σ with the vertices

$$A_1, A_2, \ldots, A_n, A_{n+1} \qquad (A_{n+1} = A_1)$$

such that the integral curvature of every arc $\gamma_i = \widehat{A_i A_{i+1}}$ of γ is not greater than π. On each arc γ_i take a point B_i where the tangent line is parallel to the straight line $A_i A_{i+1}$. Denote by α_i the inner angle of the polygonal line σ at the vertex A_i. Then $\int_{\bar\gamma_i} k(s)\, ds = \pi - \alpha_i$, where $\bar\gamma_i$ is the arc of γ from B_i to B_{i+1}. Consequently,

$$\int_\gamma k(s)\, ds = \sum_{i=1}^n \int_{\bar\gamma_i} k(s)\, ds = n\pi - \sum_{i=1}^n \alpha_i.$$

On the other hand, $\sum_{i=1}^n \alpha_i = \pi(n - 2) = n\pi - 2\pi$. Hence

$$\int_\gamma k(s)\, ds = n\pi - n\pi + 2\pi = 2\pi. \qquad\square$$

Problem 1.7.8. If γ is a closed curve whose curvature at each point is not smaller than $\frac{1}{a} > 0$, then

(1) $l(\gamma) \le 2\pi a$,   (2) the area $S(D(\gamma)) \le \pi a^2$,   (3) the diameter $d \le 2a$,

and equality holds for all the above cases if and only if γ is a circle of radius a.

Solution. From Problem 1.7.6 it follows that γ is convex. Let AB be the diameter of γ. Find on γ two points C and D where the tangent lines are parallel to AB. Drop perpendiculars CO_1 and DO_2 from the points C and D onto AB.

Prove statement (1). Take the arc BC and introduce the following coordinate system: O_1 is the origin, O_1C is the OX axis, and O_1B is the OY axis. The integral curvature of the arc CB is $\pi/2$, and hence $\int_0^{l_0} k(t)\, dt = \pi/2$, where l_0 is the arc length of BC. Since $k(t) \ge 1/a$, then $\frac{l_0}{a} \le \frac{\pi}{2}$ or $l_0 \le \pi a/2$, and equality holds if and only if $k(t) \equiv 1/a$. Analogously, $l(CA) \le \pi a/2$, $l(AD) \le \pi a/2$, and $l(DB) \le \pi a/2$. Hence $l(\gamma) \le 2\pi a$.

The second statement of the problem follows from the isoperimetric inequality

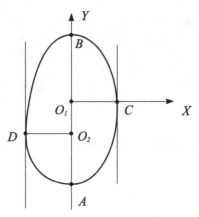

Figure 1.15. Solution of Problem 1.7.8.

$$S(D(\gamma)) \leq \frac{l^2}{4\pi} \leq \frac{4\pi^2 a^2}{4\pi} = \pi a^2.$$

Prove the third statement of the problem. Write down the equation of the arc CB of the curve γ in terms of its curvature $k(t)$ (see Theorem 1.6.3) as

$$x = x_0 + \int_0^t \cos\left[\int_0^s k(t)\, dt - \alpha_0\right] ds,$$

$$y = \int_0^{t_0} \sin\left[\int_0^s k(t)\, dt - \alpha_0\right] ds.$$

From the definition of the coordinate system it follows that

$$\alpha_0 = \frac{\pi}{2}, \qquad O_1 B = y(l_0) = \int_0^{l_0} \cos\left[\int_0^s k(t)\, dt\right] ds.$$

Since

$$\int_0^s k(t)\, dt \geq \frac{s}{a}$$

and

$$0 \leq \frac{l_0}{a} \leq \int_0^{l_0} k(t)\, dt = \frac{\pi}{2},$$

then

$$\cos\frac{s}{a} \geq \cos\left(\int_0^s k(t)\, dt\right) \quad \text{and} \quad 0 < \sin\frac{l}{a} \leq 1.$$

Thus

$$O_1 B = \int_0^{l_0} \cos\left(\int_0^s k(t)\, dt\right) ds \leq \int_0^{l_0} \cos\frac{s}{a}\, ds = a \sin\frac{l_0}{a} < a.$$

Analogously, $O_1 A < a$ and $AB = O_1 A + O_1 B < 2a$. Here equality is possible if and only if $k(t) \equiv 1/a$. $\qquad \square$

Formulate and solve the dual problem to Problem 1.7.8 for convex curves.

Problem 1.7.9 (Problem about a bent bow). Let the arcs of convex curves γ_1 and γ_2 have the same length l. Assume that their curvatures $k_1(t)$ and $k_2(t)$ obey the inequality $k_1(t) \geq k_2(t) \geq 0$ and let $\int_0^l k_1(t)\,dt < \pi$. Then $\gamma_1(0)\gamma_1(l) \leq \gamma_2(0)\gamma_2(l)$, and equality holds if and only if $k_1(t) \equiv k_2(t)$.

Solution. Find a point $\gamma_1(s_0)$ on the curve γ_1 where the tangent line to γ_1 is parallel to the chord $\gamma_1(0)\gamma_1(l)$. Draw an orthogonal coordinate system in the following way: $\gamma_1(s_0)$ is the origin, the OX axis coincides with a tangent line to γ_1, and the OY axis is orthogonal to the OX axis and directed to the chord $\gamma_1(0)\gamma_1(l)$. Translate γ_2 so that the point $\gamma_2(s_0)$ coincides with $\gamma_1(s_0)$ and the tangent line to γ_2 at a point $\gamma_2(s_0)$ coincides with the OX axis. Denote by B the point of intersection of the OY axis with the chord $\gamma_1(0)\gamma_1(l)$. The equations of the curves γ_1 and γ_2 in our coordinate system have the form

$$\gamma_1: \begin{cases} x = x_1(s) = \int_{s_0}^s \cos\left[\int_{s_0}^s k_1(t)\,dt\right] ds, \\ y = y_1(s), \end{cases}$$

$$\gamma_2: \begin{cases} x = x_2(s) = \int_0^s \cos\left[\int_{s_0}^s k_2(t)\,dt\right] ds, \\ y = y_2(s). \end{cases}$$

Then $x_1(l) = B\gamma_1(l)$, and $x_2(l)$ is equal to the orthogonal projection of the chord $\gamma_1(s_0)\gamma_2(l)$ onto the OX axis. Prove that $x_1(l) \leq x_2(l)$. Since $0 < \int_{s_0}^s k(t)\,dt < \pi$ for $s_0 < s < l$, then

$$x_1(l) = \int_{s_0}^l \cos\left[\int_{s_0}^s k_1(t)\,dt\right] ds \leq \int_{s_0}^l \cos\left[\int_{s_0}^s k_2(t)\,dt\right] ds = x_2(l).$$

Analogously, $x_1(0) = |B\gamma(0)|$ is not greater than the projection of the chord $\gamma_2(s_0)x_2(0)$ onto the OX axis. Thus $\gamma_1(0)\gamma_1(l)$ is not greater than a sum of the orthogonal projections of the chords $\gamma_2(0)\gamma_2(s_0)$ and $\gamma_2(s_0)\gamma_2(l)$, which at the same time is not greater than $\gamma_2(0)\gamma_2(l)$. Equality holds if and only if $k_1(s) \equiv k_2(s)$. □

Problem 1.7.10. If γ is a closed curve whose curvature at each point is not smaller than $1/a$, then it can be rolled without sliding inside a disk of radius a.

Solution. First, consider the case $k(t) > 1/a$. Locate a circle $C(a)$ of radius a so that the origin O belongs to $C(a)$ and the OX axis (through the point O) is tangent to $C(a)$. Take an arbitrary point P on γ and locate γ so that $P = O$ and the tangent line to γ at P coincides with the OX axis. Let P_1 be a point on γ such that the arcs γ_1 and γ_2 into which γ is divided by the point P have integral curvature π.

Introduce the arc length parameter s (counted from P) on γ_1 and $C(a)$. Then $\alpha(s) = \int_0^s k(t)\,dt$ is not greater than π. We show that $\gamma_1 \cap C(a) = \emptyset$. If not, let P_2 be the first point (starting from P) of intersection of γ_1 with $C(a)$. Write the equations of the curves γ and $C(a)$:

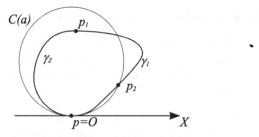

Figure 1.16. Solution of Problem 1.7.10.

$$\gamma_1: \begin{cases} x = \int_0^s \cos\alpha(s)\,ds, \\ y = y_1(s), \end{cases} \qquad C(a): \begin{cases} x = \int_0^s \cos(\frac{s}{a})\,ds, \\ y = y_2(s). \end{cases}$$

Then there exist numbers s_2 and s_1 such that

$$\int_0^{s_2} \cos\alpha(s)\,ds = \int_0^{s_1} \cos\frac{s}{a}\,ds, \qquad P_2 = \gamma(s_2) = C(a)(s_1).$$

Since $\alpha(s) = \int_0^s k(t)\,dt > \frac{s}{a}$, then $\cos\alpha(s) < \cos\frac{s}{a}$. Hence $s_2 > s_1$. But on the other hand, the convex arc PP_2 of γ_1 lies entirely inside the arc PP_2 of the circle $C(a)$ and the chord PP_2. Thus $s_2 \le s_1$, which is a contradiction. Hence $\gamma_1 \cap C(a) = \emptyset$.

Analogously, one can prove that $\gamma_2 \cap C(a) = \emptyset$. Now if $k(t) \ge 1/a$, then from the above, it follows that $\gamma \cap C(a+\varepsilon) = \emptyset$ for every $\varepsilon > 0$. From this we get that γ lies entirely inside of $C(a)$. The problem is solved in view of the arbitrariness of the point P.   □

Formulate and solve the dual problem to Problem 1.7.10 for convex curves.

Problem 1.7.11. Let a curve γ touch the circle $C(a)$ with center O and radius a at the points A and B and lie entirely inside of $C(a)$, and suppose $\angle AOB < \pi$. Then the curvature at some point on γ is smaller than $1/a$.

Solution. Consider a curve $\bar{\gamma}$, composed from the greater circular arc of $C(a)$ and of a curve γ. Assume that the curvature at all points of γ is not smaller than $1/a$.

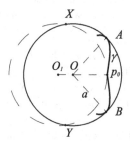

Figure 1.17. Solution of Problem 1.7.11.

Let P_0 be a point on γ nearest to the center O of a circle $C(a)$. Then OP_0, by conditions of the problem, is smaller than a. Denote by O_1 a center of a circle of a radius a, which touches $\bar{\gamma}$ at P_0. Then this circle intersects $\overline{\gamma}$, in contradiction to the statement of Problem 1.7.10.   □

Problem 1.7.12. Let a curve γ touch a circle $C(a)$ of radius a at the points A and B, located outside of $C(a)$, and suppose $\angle AOB < \pi$. Prove that there is a point on γ at which the curvature of γ is greater than $1/a$.

Prove this problem on your own.

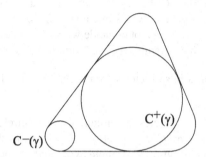

Figure 1.18. Solution of Problem 1.7.12.

In order to formulate the next problems, we give some definitions.

Let γ be any smooth closed convex curve. Denote by $C(P, \gamma)$ the circle satisfying the following conditions:

(1) $C(P, \gamma)$ *touches* γ *at the point* P,

(2) $C(P, \gamma) \subset D(\gamma)$,

(3) $C(P, \gamma)$ *has the maximal radius for which conditions* (1) *and* (2) *hold.*

Denote by $C^+(\gamma)$ a circle of maximal radius contained in $D(\gamma)$. Let $C^-(\gamma)$ be a circle of minimal radius satisfying the conditions (1)–(3) (for some $P \in \gamma$). Denote by $R(P, \gamma)$ the radius of $C(P, \gamma)$, and then write the radii of the circles $C^+(\gamma)$ and $C^-(\gamma)$, respectively, as

$$R^+(\gamma) = \sup_{P \in \gamma} R(P, \gamma), \qquad R^-(\gamma) = \inf_{P \in \gamma} R(P, \gamma).$$

Problem 1.7.13. If $C(P, \gamma) \cap \gamma = P$, then the curvatures of γ and $C(P, \gamma)$ at the point P are equal.

Solution. In view of Problem 1.7.1, the curvature $k_\gamma(P)$ of the curve γ at P is not greater than $1/R(P, \gamma)$. Assume that $k_\gamma(P) < 1/R(P, \gamma)$. Take a monotonic sequence of numbers R_n satisfying the conditions

$$k_\gamma(P) < \frac{1}{R_n} < \frac{1}{R(P, \gamma)}, \qquad \lim_{n \to \infty} R_n = R(P, \gamma).$$

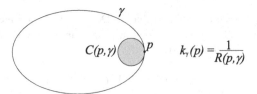

Figure 1.19. Solution of Problem 1.7.13.

Denote by $C(P, \gamma, R_n)$ the circle of radius R_n that touches γ at P, in view of the definition of $C(P, \gamma)$. Then $C(P, \gamma, R_n)$ intersects γ at least at one more point $P_n \neq P$. Without loss of generality, assume that $\lim_{n \to \infty} P_n = \bar{P}$ exists. In view of Problem 1.7.1, the point $\bar{P}$ cannot coincide with P. Consequently, $\bar{P} \neq P$ and $\bar{P} \in \gamma \cap C(P, \gamma)$, which contradicts the condition of the problem.   □

Problem 1.7.14. On an arbitrary closed semicircle of a circle $C^+(\gamma)$ there is a point that belongs to γ.

Solution. If not, let $\bar{C}$ be the closed semicircle of the circle $C^+(\gamma)$ for which $\bar{C} \cap \gamma = \emptyset$ holds. Denote by O the center of $C^+(\gamma)$, and by R^+ its radius. Draw the diameter a through the ends of $\bar{C}$. Denote by A_1 and A_2 the intersection points of the straight line containing the diameter a with γ. The points A_1 and A_2 divide γ into two arcs γ_1 and γ_2. Denote by γ_1 one of these arcs satisfying the property $\gamma_1 \cap C^+(\gamma) = \emptyset$. Let $d_0 = \min_{\gamma(s) \in \gamma_1} (O\gamma(s) - R^+) > 0$. Drop the perpendicular

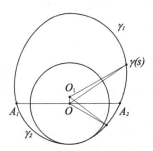

Figure 1.20. Solution of Problem 1.7.14.

to the line a from the center O inside of the semicircle $\bar{C}$ and then mark off the line segment $OO_1 = d_0/2 = \delta$ on it. For $\gamma(s) \in \gamma_1$.

$$O_1\gamma(s) - R^+ > O\gamma(s) - \frac{d_0}{2} - R^+ \geq \delta > 0.$$

For $\gamma(s) \in \gamma_2$.

$$O_1\gamma(s) - R^+ = \sqrt{(O\gamma(s))^2 + \delta^2 - 2O\gamma(s) \cdot \delta \cos\alpha} - R^+,$$

where $\alpha(s)$ is the angle in $\triangle OO_1\gamma(s)$ at the vertex O. From the definition of O_1 it follows that $\alpha(s) \geq \pi/2$. Hence

$$O_1\gamma(s) - R^+ \geq \sqrt{(O\gamma(s))^2 + \delta^2} - R^+ \geq \sqrt{(R^+)^2 + \delta^2} - R^+$$

$$= \frac{\delta^2}{\sqrt{(R^+)^2 + \delta^2} + R^+} = \delta_1 > 0.$$

But then the circle with center at O_1 and radius $R = R^+ + \sigma > R^+$ lies entirely inside of $D(\gamma)$, in contradiction to the definition of $C^+(\gamma)$. Here $\sigma = \frac{1}{2}\min(\delta, \delta_1)$. The reader should consider the case in which $C^+(\gamma) \cap \gamma$ consists of exactly two points. $\qquad\square$

Problem 1.7.15. If γ is a closed convex curve of class C^2, then the set $C_0 = C^-(\gamma) \cap \gamma$ is connected.

Solution. C_0 is a closed set. Consequently, the set $\gamma \setminus C_0$ is the union of open arcs. If C_0 is not connected, the number of these arcs $\gamma_1, \gamma_2, \ldots, \gamma_k$ is not smaller than 2.

Two cases are possible:

(1) The integral curvature of one of these arcs is smaller than π.
(2) C_0 consists of two diametrically opposite points P_1 and P_2 of the circle $C^-(\gamma)$.

In the first case there is a point P_1 on the arc γ_1 at which the curvature $k_\gamma(P_1)$ is greater than $1/R^-(\gamma)$, but then $R(P_1, \gamma) < R^-(\gamma)$, in contradiction to the definition of $C^-(\gamma)$.

In the second case, for any point $P \in \gamma$, $P \notin C_0$, the value $R(P, \gamma)$ is not greater than $R^-(\gamma)$, because $C(P, \gamma)$ belongs, together with its curve γ, to the strip of width $2R^-(\gamma)$ formed by the tangent lines to γ at the points P_1 and P_2. Hence, in this case the equality $R(P, \gamma) = R^-(\gamma)$ holds for all P. But then $\gamma = \bar{C}(\gamma)$, in contradiction to the assumption of nonconnectedness. $\qquad\square$

Formulate and solve the dual problems to Problems 1.7.14 and 1.7.15.

Problem 1.7.16 (The four-vertex theorem).[1] Prove that for any closed convex curve of class C^2 the curvature function $k(s)$ has at least two local minima and two local maxima.

Note that a closed convex curve is sometimes called an *oval*. The converse of the four-vertex theorem is also true (H. Gluck, 1971): *a function on a circle can be realized as the curvature of a closed plane curve exactly if it admits at least two maxima separated by two minima.*

Solution. If $C^+(\gamma) \neq \gamma$, then $C_1 = C^+(\gamma) \cap \gamma$ divides $\gamma \setminus C_1$ into open arcs $\gamma_1, \gamma_2, \ldots, \gamma_k$, $k \geq 2$. Moreover, either the integral curvature of at least two arcs,

---

[1] There are various recent generalizations of four vertex theorem. See, for example, Sedykh, V.D., *The four-vertex theorem of a plane curve and its generalizations.* (Russian. English summary). Soros. Obraz. Zh., Vol. 6, No. 9, 122–127, 2000; Tabachnikov, S. *A four vertex theorem for polygons.* Am. Math. Mon., vol. 107, No. 9, 830–833, 2000.

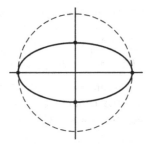

Figure 1.21. Four vertices on an ellipse.

say γ_1 and γ_2, is smaller than π, or $C_1 = \{P_1, P_2\}$, where P_1 and P_2 are diametrically opposite points on $C^+(\gamma)$.

In the first case, on each of the arcs γ_1 and γ_2, in view of Problem 1.7.11, there is at least one maximum, and between them there are at least two minima.

In the second case, consider the curve formed by the arc γ_1 and the arc of the circle $C^+(\gamma)$. If for this curve $C^-(\gamma)$ does not coincide with $C^+(\gamma)$, then at the point of tangency of $C^-(\gamma) \subset \gamma_1$, the curvature of γ is greater than the curvature of γ at the points P_1 and P_2. Consequently, there is a local maximum of the curvature on the arc γ_1. An analogous statement holds for the arc γ_2. Again we obtain at least two maxima and between them at least two minima. But if $C^+(\gamma) = C^-(\gamma)$, then γ is a circle.   □

Another proof of the four-vertex theorem is based on Hurwitz's theorem.

Theorem 1.7.1 (Hurwitz). *Let a continuous function $f(\varphi)$ have period 2π and be orthogonal to cosine and sine in that*

$$\int_0^{2\pi} f(\varphi) \sin \varphi \, d\varphi = 0, \qquad \int_0^{2\pi} f(\varphi) \cos \varphi \, d\varphi = 0.$$

Then on the closed interval $[0, 2\pi]$ it has at least two local minima and two local maxima.

Proof (suggested by V.V. Ivanov). First, note that if on the interval of the period, a function has, say, two points of local minimum, then it necessarily has two points of local maximum. Indeed, two points of minimum marked on a circle divide it into two arcs. Obviously, strictly inside of each of these arcs there is at least one local maximum of $f(\varphi)$. The same arguments provide that if a function has two points of maximum, then it also has two points of minimum.

Owing to continuity and periodicity, the function $f(\varphi)$ necessarily takes maximum and minimum values on this closed interval. Hence, we need to find in the same interval only one more point of extremum. Assume the opposite; i.e., $f(\varphi)$ has no such points. Select the initial point on a circle so that for $\varphi = 0$, and then also for $\varphi = 2\pi$, the function $f(\varphi)$ reaches its maximum, and find a point $\varphi_0 \in (0, 2\pi)$ where our function has a minimum. Then, obviously, it monotonically increases on the closed interval $0 \le \varphi \le \varphi_0$ and strongly decreases on the closed interval $\varphi_0 \leq \varphi \leq 2\pi$. It is easy to see that under these circumstances the integrals are negative:

$$\int_0^{\varphi_0} f(\varphi) \sin\left(\varphi - \frac{\varphi_0}{2}\right) d\varphi < 0, \qquad \int_{\varphi_0}^{2\pi} f(\varphi) \sin\left(\varphi - \frac{\varphi_0}{2}\right) d\varphi < 0.$$

Indeed, the sine term figured in them is symmetric on both intervals of integration relative to their midpoints $\varphi_0/2$ and $\pi + \varphi_0/2$; furthermore, on the region, where the sine is positive the values of a function $f(\varphi)$ are smaller than on the region, where the sine is negative.

It is not difficult to translate these "geometric" arguments into the exact language of formulas. For example, the first of above integrals can be written in the form

$$\int_0^{\varphi_0/2} \left[f\left(\varphi + \frac{\varphi_0}{2}\right) - f\left(\varphi - \frac{\varphi_0}{2}\right) \right] \sin \varphi \, d\varphi,$$

and we see clearly that it is negative. The second of the above integrals is transformed analogously. Thus, the sum of above two integrals is also negative;

$$\int_0^{2\pi} f(\varphi) \sin\left(\varphi - \frac{\varphi_0}{2}\right) d\varphi < 0,$$

but by the condition it must be zero.   □

Here is a second solution of Problem 1.7.16.

Proof (2). Assume without loss of generality that the curvature of the oval is positive at each point. If L denotes the length of an oval and s is selected as an arc length parameter, then in Cartesian coordinates (x, y) our curve can be described by the equations

$$x = x(s), \quad y = y(s), \qquad 0 \leq s \leq L.$$

From Theorem 1.6.3 it follows that

$$x'(s) = \cos \varphi(s), \qquad y'(s) = \sin \varphi(s), \qquad 0 \leq \varphi(s) < 2\pi,$$

and let $\varphi(0) = 0$. The curvature of the oval is $k(s) = \varphi'(s)$ (see Theorem 1.6.3). Since the curvature is positive, the angle $\varphi = \varphi(s)$ is an increasing function of the variable s, and hence an inverse function $s = s(\varphi)$ exists. Writing for simplicity $k(\varphi) = k(s(\varphi))$, we see that the curvature k is a continuous 2π-periodic function of φ. Note also that a function $s = s(\varphi)$ has a derivative at each point φ, which is calculated by the formula

$$\frac{ds}{d\varphi} = \frac{1}{\varphi'(s(\varphi))} = \frac{1}{k(\varphi)}.$$

From this follows $ds = \frac{d\varphi}{k(\varphi)}$. Now we are ready to present the solution of Problem 1.7.16.

The function $1/k(\varphi)$ is orthogonal (in the integral sense) to both the cosine and the sine of φ. Indeed, an oval is a closed curve, and hence

$$\int_0^L x'(s)\,ds = \int_0^L y'(s)\,ds = 0.$$

But on the other hand, these integrals are

$$\int_0^L \cos\varphi(s)\,ds = \int_0^{2\pi} \cos\varphi\,\frac{d\varphi}{k(\varphi)} \quad \text{and} \quad \int_0^L \sin\varphi(s)\,ds = \int_0^{2\pi} \sin\varphi\,\frac{d\varphi}{k(\varphi)}.$$

For finishing the proof, note that the functions $k(\varphi)$ and $1/k(\varphi)$ have a common extremum, and refer to Hurwitz's theorem, Theorem 1.7.1.   □

We shall formulate one more problem, whose solution (suggested by V.V. Ivanov) is also based on Hurwitz's theorem.

Imagine a convex polygon contained in a vertical plane. Call each of the sides on which it can lean and remain on a horizontal straight line without overturning under the action of gravity a *base of the polygon*. A base of a polygon is *stable* if the polygon standing on it would not drop under small rotations in the vertical plane in one or another direction. Clearly, at least one such base exists. But the situation is much more interesting.

Problem 1.7.17 (The bases of a convex polygon). Any convex polygon has at least two stable bases.

Proof. Introduce Cartesian coordinates (x, y) in the plane of a convex polygon M, taking as origin the "center of mass" of our convex figure, which, obviously, is located strictly inside of M. Note first that

$$\iint_M x\,dx\,dy = 0, \qquad \iint_M y\,dx\,dy = 0,$$

and second, we may describe the boundary of M in terms of corresponding polar coordinates (ρ, φ) by an equation of the form $\rho = \rho(\varphi)$, where the polar radius ρ constitutes a positive continuous function of the angular variable φ running from 0 to 2π. Recomputing the above double integrals in polar coordinates, we easily obtain two new equalities,

$$\int_0^{2\pi} \varrho^3(\varphi)\cos\varphi\,d\varphi = 0, \qquad \int_0^{2\pi} \varrho^3(\varphi)\sin\varphi\,d\varphi = 0.$$

Hence, by Hurwitz's theorem, the cube of the polar radius, and so the polar radius itself, has two minima. From elementary geometrical considerations it is clear that these two values of the polar radius show us in what directions the rays from the origin to the sides of a polygon must be drawn in order to intersect the interiors of these sides exactly in a right angle. Such sides of the polygon will be the required stable bases.   □

As we see from the solution of the problem, it remains true for arbitrary bounded convex regions if together with "bases" we speak about stable "points of support."

Problem 1.7.18. Prove that if there is a circle intersecting an oval at $2n$ points, then there exist $2n$ vertices on this oval.

Hint. Use Problems 1.7.16 and 1.7.17.

Problem 1.7.19. Let γ be a simple regular closed curve of class C^2. Denote by $D(\gamma)$ the region bounded by γ. Prove that if the curvature $k(P)$ of γ at each point $P \in \gamma$ satisfies the inequality $|k(P)| \leq 1/a$, then there is a disk of radius a lying entirely in $D(\gamma)$.

Problem 1.7.19 was formulated by A. Fet and solved by V. Ionin.

Hint. For solving the problem one needs to consider and study the properties of the "central" set of a curve γ. The *central set* M of the region $D(\gamma)$ (a curve γ) consists of the points defined in the following way: Let $Q \in \gamma$. Denote by $C(Q)$ the disk of a maximal radius that touches γ at the point Q and inside $D(\gamma)$. The set of the centers of the disks $C(Q)$, when Q runs along the whole curve γ, forms the set M.

Problem 1.7.20. The integral curvature of an infinite convex curve γ is not greater than π.

Solution. Let $\gamma(s)$ be the arc length parameterization of a curve γ, counted from one of its points. We show that for any s_1 and s_2, the inequality $\int_{s_1}^{s_2} k(s)\,ds \leq \pi$ holds.

If not, let s_1 and s_2 be numbers such that $\int_{s_1}^{s_2} k(s)\,ds = \omega_1 > \pi$. Draw the straight lines a_1 and a_2 that touch γ at the points $\gamma(s_1)$ and $\gamma(s_2)$, respectively. Since a_1 and a_2 are nonparallel, they intersect at some point A. Thus the region D bounded by the line segments $\gamma(s_2)A$, $\gamma(s_1)A$, and the arc $\widehat{\gamma(s_1)\gamma(s_2)}$ of γ that is defined by the inequalities $s_1 \leq s \leq s_2$ is a convex compact region containing the whole curve γ, contradicts the condition of the problem. Consequently, for any s_1 and s_2, the inequality $\int_{s_1}^{s_2} k(s)\,ds \leq \pi$ holds. $\qquad\square$

Problem 1.7.21. If the curvature function $k(s)$ of a curve $\gamma(s)$ $(-\infty < s < \infty)$ is a positive strictly increasing function, then γ has no points of self-intersection. Here s is an arc length parameter starting from some point of γ.

Solution. If not, let s_1 and s_2 be real numbers such that $\gamma(s_1) = \gamma(s_2)$ and the arc $\sigma = \widehat{\gamma(s_1)\gamma(s_2)}$ of $\gamma(s)$ has no other points of self-intersection for $s_1 \leq s \leq s_2$. Then the curve σ bounds some convex region D. Let $C(0, R)$ be a disk of a maximal radius inscribed in D, the point O be its center, and R its radius.

From Problem 1.7.12 it follows that the circle $C(0, R)$ touches σ at least at two points $\gamma(s_3)$ and $\gamma(s_4)$, where $s_1 < s_3 < s_2$ and $s_1 < s_4 < s_2$, and $\angle\gamma(s_3)O\gamma(s_4)$ is not greater than π. But then from Problems 1.7.1 and 1.7.10 it follows that at

$\gamma(s_3)$ and $\gamma(s_4)$, the curvature of σ is not greater than $1/R$, and there is an inner point on the arc $\gamma\overgroup{(s_3)\gamma(s_4)}$ of the curve σ for which the curvature is greater than $1/R$. This is a contradiction, because $k(s)$ is an increasing function.   □

Problem 1.7.22. Under the conditions of Problem 1.7.19,

$$\lim_{s\to\infty} k(s) = a, \qquad \lim_{s\to-\infty} k(s) = b$$

(the cases $a = 0$ and $b = \infty$ are not excluded). Then there exist circles $C_1(1/b)$ and $C_2(1/a)$ with radii $1/b$ and $1/a$, respectively, such that a curve γ winds in a spiral onto $C_1(1/b)$ from outside, and onto $C_2(1/a)$ from inside.

Hint. Solve on your own. Note that for $a = 0$ the curve γ has an asymptote, and for $b = \infty$ it winds in a spiral onto a point.

1.7.1 Parallel Curves

Let $\gamma(t)$ be a smooth regular parameterization of a curve γ, and let $\vec{e}(t)$ be a continuous vector field of unit normals along γ, $a \le t \le b$. Draw a set γ_d by marking from each point $\gamma(t)$ the line segment of length $|d|$ in the direction of $\vec{e}(t)$ if $d > 0$, and in the direction $-\vec{e}(t)$ if $d < 0$. A set γ_d is called a *parallel curve* corresponding to the curve γ. This set is not necessarily a curve. For example, a curve γ_a parallel to a circle of radius a is a point. Moreover, γ_d is not necessarily a regular curve at all its points. The regularity conditions of γ_d are formulated below, in Theorem 1.7.2. It is easy to obtain the equations of γ_d if the equations of γ are known. Let $x = x(t)$, $y = y(t)$ be the equations of γ, and t an arc length parameter, $a \le t \le b$. Then the equations of γ_d are

$$\begin{cases} x = x_d(t) = x(t) \pm y'(t)d, \\ y = y_d(t) = y(t) \mp x'(t)d, \end{cases} \tag{1.21}$$

where the signs $(+, -)$ or $(-, +)$ depend on the choice of direction of the vector field $\vec{e}(t)$. Define the curvature signs of γ and γ_d with the help of the vector field $\vec{e}(t)$.

Theorem 1.7.2. *If a curve γ with parameterization $\gamma(t)$, $a \le t \le b$ ($a = -\infty$, $b = \infty$ allowed), is regular of class C^2, and for all $t \in [a, b]$ the inequality $d \ne 1/k(t)$ holds, then the parallel curve γ_d is regular, and its curvature $k_d(t)$ is related to the curvature $k(t)$ of γ by the formula*

$$k_d(t) = \frac{k(t)}{1 - k(t)d}.$$

Proof. Let $x = x(t)$ and $y = y(t)$ be the equations of γ and assume that t is the arc length parameter, and $\vec{e}(t)$ coincides with the vector field $(-y'(t), x'(t))$. Then the equations of γ_d take the form

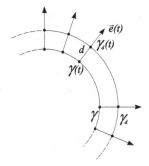

Figure 1.22. Parallel curves.

$$\begin{cases} x = x_d(t) = x(t) - y'(t)d, \\ y = y_d(t) = y(t) + x'(t)d. \end{cases}$$

Consequently,

$$\begin{cases} x_d'(t) = x'(t) - y''(t)d, \\ y_d'(t) = y'(t) + x''(t)d. \end{cases}$$

From the rule for the choice of the sign of the curvature $k(t)$ of γ, it follows that

$$x''(t) = -k(t)y', \qquad y''(t) = k(t)x'(t).$$

Indeed, if the components $\vec{v}_1(t)$ and $\vec{v}_2(t)$ of the principal normal vector $\vec{v}$ of γ are given by the equalities $\vec{v}_1 = -y'$, $\vec{v}_2 = x'$, then the curvature $k(t)$ of γ at the point $\gamma(t)$ is positive, and from Frenet formulas we obtain

$$x'' = -|k|y' = -ky', \qquad y'' = |k|x' = kx'.$$

If $\vec{v}_1 = y'$, $\vec{v}_2 = -x'$ hold, then the curvature $k(t)$ of γ at the point $\gamma(t)$ is negative, and again from the Frenet formulas we obtain

$$x'' = |k|y' = -ky', \qquad y'' = -|k|x' = kx'.$$

Finally, we have

$$\begin{cases} x_d' = x' - y''d = x'(1 - kd), \\ y_d' = y' + x''d = y'(1 - kd). \end{cases} \tag{1.22}$$

From this it follows that

$$(x_d')^2 + (y_d')^2 = (1 - kd)^2(x'^2 + y'^2) = (1 - kd)^2 \neq 0$$

in view of condition of the theorem. The first statement of the theorem is proven.

Now find the curvature $k_d(t)$ of the parallel curve γ_d at a point $\gamma_d(t)$. Take a point $\gamma_d(t + \Delta t)$ and find the angle $\Delta\varphi(t)$ between the tangent lines to γ_d at the points $\gamma_d(t)$ and $\gamma_d(t + \Delta t)$. As is seen from formulas (1.22), the angle $\Delta\varphi(t)$

is equal to the angle $\Delta\theta(t)$ between the tangent lines to γ at the points $\gamma(t)$ and $\gamma(t + \Delta t)$:

$$\Delta\varphi(t) = \Delta\theta(t). \tag{1.23}$$

The arc length Δs of a curve γ_d between the points $\gamma_d(t)$ and $\gamma_d(t + \Delta t)$ is given by the formula

$$\Delta s = \int_t^{t+\Delta t} |1 - kd| dt. \tag{1.24}$$

So from (1.23) and (1.24) it follows that

$$|k_d(t)| = \lim_{\Delta t \to 0} \frac{\Delta\varphi(t)}{\Delta s} = \lim_{\Delta t \to 0} \frac{\Delta\theta(t)}{\Delta t} \cdot \lim_{\Delta t \to 0} \frac{\Delta t}{\Delta s} = \frac{|k(t)|}{|1 - kd|}. \tag{1.25}$$

From the definition of the signs of the curvatures $k(t)$ and $k_d(t)$ it is not difficult to deduce that for $1 - kd > 0$ the signs of $k(t)$ and $k_d(t)$ coincide, but for $1 - kd < 0$ these signs are opposite. Hence from (1.25) the second statement of the theorem follows.   □

Remark 1.7.1. If we define $R(t)$ and $R_d(t)$ by the formulas

$$R(t) = \frac{1}{k(t)}, \qquad R_d(t) = \frac{1}{k_d(t)},$$

then the last statement of Theorem 1.7.2 takes the form $R_d = R - d$.

Problem 1.7.23. Let $\gamma(t)$ $(-\infty < a \le t \le b < \infty)$ be a regular parameterization of a curve γ of the class C^2. Prove that if

$$|d| < \inf_{t \in [a,b]} \frac{1}{k(t)},$$

then $\gamma_d \cup \gamma_{-d}$ can be defined as a set of points whose distances from γ are equal to $|d|$.

Hint. If not, use Theorem 1.7.2 to obtain a contradiction.

Problem 1.7.24. Find a smooth regular curve, for which the parallel curves γ_d have nonregular points for every d.

1.7.2 Evolutes and Evolvents

For regular curves of class C^2 one may define a curve called an evolute of a given curve γ.

Let $\gamma(t)$ $(a \le t \le b)$ have the property $k(t) \ne 0$ for all $t \in [a, b]$. Then at each point of γ the principal normal vector $\vec{v}(t)$ is defined. Mark off the line segment

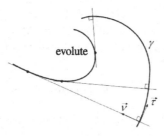

Figure 1.23. Evolute of a curve.

of the length $R(t) = 1/k(t)$ from each point $\gamma(t)$ in the direction of $\vec{v}(t)$. The ends of these line segments form a set of points called the *evolute* of γ.

The evolute of a curve is not necessarily a curve. So for example, the evolute of a circle is a point. From Theorem 1.7.2 it follows that the locus of singular (not regular) points for the parallel curves is the evolute.

In the general case, when the condition $k(t) \neq 0$ is not satisfied at all points of γ, we may define the evolute of a curve γ independently for each of its arcs where the condition $k(t) \neq 0$ holds.

If the equations of a curve γ are given by the arc length parameterization $x = x(t)$, $y = y(t)$, then the equations of the evolute are written in the following form:

$$\begin{cases} x = \tilde{x}(t) = x(t) + R(t)\vec{v}_1(t), \\ y = \tilde{y}(t) = y(t) + R(t)\vec{v}_2(t), \end{cases} \tag{1.26}$$

where $\vec{v}_1(t)$ and $\vec{v}_2(t)$ are the components of the principal normal vector $\vec{v}(t)$. Find the condition for a given parameterization of an evolute to be regular. Differentiating (1.26) with respect to t, we obtain

$$\begin{cases} \tilde{x}'(t) = x'(t) + R(t)\vec{v}_1'(t) + R'(t)\vec{v}_1(t), \\ \tilde{y}'(t) = y'(t) + R(t)\vec{v}_2'(t) + R'(t)\vec{v}_2(t). \end{cases} \tag{1.27}$$

By the Frenet formulas, $\vec{v}_1'(t) = -kx'(t)$, $\vec{v}_2'(t) = ky'(t)$. Hence

$$\begin{cases} \tilde{x}'(t) = R'(t)\vec{v}_1(t), \\ \tilde{y}'(t) = R'(t)\vec{v}_2(t). \end{cases} \tag{1.28}$$

From (1.28) it follows that $(\tilde{x}')^2 + (\tilde{y}')^2 = (R')^2$. Consequently, if $R'(t) \neq 0$, then $(\tilde{x}(t), \tilde{y}(t))$ is a regular point of the evolute. From equalities (1.28) the main property of the evolute follows: *a tangent line to the evolute at a point $(\tilde{x}(t), \tilde{y}(t))$ is a normal line to the curve γ at the point $(x(t), y(t))$.*

So, the following picture is obtained.

1. If along the arc $\gamma(t)$ ($a \le t \le b$), $k(t) \neq 0$ and $k'(t) \neq 0$ hold then an evolute is a regular curve.
2. At a point $\gamma(t_0)$ where $k(t_0) = 0$, a normal line to γ is the asymptote to both branches of an evolute.

3. If at a point $\gamma(t_0)$ we have $k(t) \neq 0$, but $k'(t) = 0$ and $k''(t) \neq 0$, then $(\tilde{x}(t_0), \tilde{y}(t_0))$ is a singular point of an evolute. At this point both regular arcs of the evolute meet: they have a common tangent line and are located in opposite half-planes from it.

The case $k'(t_0) = k''(t_0) = 0$ requires further investigation and is not treated here.

Now calculate the arc length s of the evolute between points with the parameters t_1 and t_2,

$$s = \int_{t_1}^{t_2} \sqrt{\tilde{x}'^2 + \tilde{y}'^2} dt = \int_{t_1}^{t_2} R'(t) \, dt = R(t_2) - R(t_1);$$

i.e., the arc length is equal to the difference in the radii of curvature of γ at the points $\gamma(t_2)$ and $\gamma(t_1)$.

Example 1.7.1. The evolute of the parabola $x = t$, $y = at^2$ is the *half-cubic (Neil's) parabola* $x_{ev} = -4a^2t^3$, $y_{ev} = 3at^2 + \frac{1}{2a}$. Its implicit equation is $27x^2 = 16a\left(y - \frac{1}{2a}\right)^3$.

Definition 1.7.1. The *evolvent* of a curve γ is a curve $\bar{\gamma}$ such that its evolute is γ.

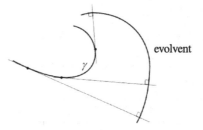

evolvent

Figure 1.24. Evolvent of a curve.

Let $x = x(t)$, $y = y(t)$ be the arc length parameterization of γ. Then the equations of the evolvent $\bar{\gamma}$ may be written in the form

$$\begin{cases} x = \bar{x}(t) = x(t) + a(t)x'(t), \\ y = \bar{y}(t) = y(t) + a(t)y'(t). \end{cases} \qquad (1.29)$$

Find the function $a(t)$ from the orthogonality of the vectors $(x'(t), y'(t))$ and $(\tilde{x}'(t), \tilde{y}'(y))$. Since

$$\tilde{x}'(t) = x'(t) + a(t)x''(t) + a'(t)x'(t), \quad \tilde{y}'(t) = y'(t) + a(t)y''(t) + a'(t)y'(t),$$

we obtain

$$x'^2(1 + a') + y'^2(1 + a') = 0,$$

or

$$a'(t) = -1,$$

from which it follows that $a(t) = -t + C$. So, the equations of an evolvent $\tilde{\gamma}(t)$ are the following:

$$\tilde{x}(t) = x(t) + x'(t)(C - t), \qquad \tilde{y}(t) = y(t) + y'(t)(C - t).$$

This means that for a given curve γ there is a whole family of evolvents depending on a constant C. Visually, the construction of an evolvent can be described by the following method. First, let a nontensional thread wind around on a part of a curve. If then the thread is unwound (maintaining tension), its endpoint's trajectory forms an evolute.

1.7.3 Curves of Constant Width

Let γ be a convex closed smooth curve. Draw a tangent line a to it through some point $Q \in \gamma$. Since the curve γ is convex and closed, there is a unique straight line $\bar{a}$ that is parallel to a but different from it and that is tangent to γ. The whole curve γ is located between these straight lines. Thus a real number $d(a)$ equal to the distance between a and $\bar{a}$ is called the *width of the curve* γ in the direction of the straight line a.

A convex closed smooth curve γ is called a *curve of constant width d* if its width does not depend on the direction of the straight line a: $d(a) = d$. Here is one example of a curve of constant width that differs from a circle.

Let $\triangle A_1 A_2 A_3$ be an equilateral triangle with a side a: $A_1 A_2 = A_1 A_3 = A_2 A_3 = a$. Draw two circles with centers at A_1, and radii r and $r + a$, and then take their arcs σ_1 and $\bar{\sigma}_1$ located outside of $\triangle A_1 A_2 A_3$ between the straight lines $A_1 A_3$ and $A_1 A_2$. Analogously, the arcs σ_2, $\bar{\sigma}_2$ and σ_3, $\bar{\sigma}_3$ are defined. The

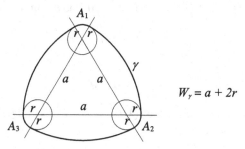

$$W_\gamma = a + 2r$$

Figure 1.25. Curve of constant width.

union $\sigma_1 \bar{\sigma}_2 \sigma_3 \bar{\sigma}_1 \sigma_2 \bar{\sigma}_3$ of these arcs forms a smooth convex curve of constant width $a + 2r$. It is also possible to define the notion of a curve of constant width for piecewise smooth convex curves if the tangent lines are replaced by supported lines. An example of such a curve is a *Rello triangle*.[2]

---

[2] For recent investigations about curves and bodies of constant width see V. Boltyanski, H. Martini and P.S. Soltan, *Excursions into Combinatorial Geometry*. Universitext. Berlin: Springer, 1997.

It may be constructed by the following method: draw a circle of radius a with the center at the vertex A_1 of $\triangle A_1 A_2 A_3$, and take its smaller arc σ_1 between A_2 and A_3. Analogously the arcs σ_2 and σ_3 are drawn. The union of arcs $\sigma_1 \sigma_2 \sigma_3$ forms a convex closed curve of constant width a. The points A_1, A_2 and A_3 are the vertices of this curve, see Figure 1.26.

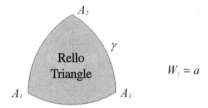

Figure 1.26. A Rello triangle.

Problem 1.7.25. Prove that the length of a curve of constant width a is πa.

Hint. Use the result of Problem 1.5.2.

Problem 1.7.26. Prove that if γ is a smooth curve of constant width b, and Q_1 and Q_2 are the points where the straight lines a_1 and a_2 touch γ, then the line segment $Q_1 Q_2$ is orthogonal to a_1 and a_2, and consequently, $Q_1 Q_2 = b$.

Solution. Let P_1 and P_2 be the points of γ for which the length of the line segment $P_1 P_2$ is equal to the diameter d of γ. Then the tangent lines to γ at the endpoints of this diameter are orthogonal to the line segment $P_1 P_2$. Consequently, $d = b$. Now let the points Q_1 and Q_2 have the properties given in the conditions of the problem. Then

$$Q_1 Q_2 \le d = P_1 P_2 = b,$$

but also

$$Q_1 Q_2 \ge b = d.$$

From these inequalities it follows that $Q_1 Q_2 = b$, and hence $Q_1 Q_2$ is orthogonal to a_1 and a_2.   □

From the statements of the last problem it follows that at each point $Q \in \gamma$ there is a unique point Q^* such that the tangent lines to γ at Q and Q^* are parallel. The points Q and Q^* are called *diametrically opposite* points on the curve γ.

Problem 1.7.27. If γ is a curve of class C^2 of the constant width a and the curvatures of γ at diametrically opposite points are equal, then γ is a circle of diameter a.

Hint. Use the result of Theorem 1.7.2.

Problem 1.7.28. Find analytic functions h such that the curve defined by the equations

$$x = h(\theta) \cos \theta - \frac{dh}{d\theta} \sin \theta, \qquad y = h(\theta) \sin \theta + \frac{dh}{d\theta} \cos \theta,$$

is a curve of constant width.

Hint. One of these functions is $h(\theta) = a + b \cos(3\theta)$ $(0 < 8b < a)$, where a and b are real constants.

1.8 Torsion of a Curve

Let a curve γ be of class C^2 and have nonzero curvature at a point P_1. Then, by continuity, the curvature of γ is nonzero on some neighborhood of P_1. Take an arbitrary point P_2 in this neighborhood. By Theorem 1.6.2 there exist unique osculating planes α_1 and α_2 at the points P_1 and P_2. Denote by $\Delta\theta$ the angle between them, and by Δs the length of the arc $P_1 P_2$ of γ.

Definition 1.8.1. The value

$$\kappa = \lim_{P_2 \to P_1} \frac{\Delta\theta}{\Delta s} = \lim_{\Delta s \to 0} \frac{\Delta\theta}{\Delta s},$$

if it exists, is called the *absolute torsion* of the curve γ at the point P_1.

If γ is a plane curve, then $\Delta\theta \equiv 0$ and the absolute torsion $\kappa = 0$. Below, we shall also prove the converse statement: *if a curve γ has nonzero curvature at each of its points and it has zero torsion, then γ is a plane curve.*

The following theorem gives us sufficient conditions of the existence of torsion and a formula to derive it.

Theorem 1.8.1. *If $\gamma: \vec{r} = \vec{r}(t)$ is a regular curve of class C^3, then at each of its points with nonzero curvature there is an absolute torsion κ, and*

$$\kappa = \frac{|(\vec{r}' \cdot \vec{r}'' \cdot \vec{r}''')|}{|\vec{r}' \times \vec{r}''|}.$$

Proof. Let $\vec{r} = \vec{r}(s)$ be the arc length parameterization of the curve. Assume that the curvature of γ at the point $P_1 = \vec{r}(s_1)$ is nonzero. Then there is a real number $\varepsilon > 0$ such that for $s \in (s_1 - \varepsilon, s_1 + \varepsilon)$ the curvature of γ at the points $\vec{r}(s)$ is also nonzero. The angle between the osculating planes at the points $P_1 = \vec{r}(s_1)$ and $P_2 = \vec{r}(s_2)$ for $s_2 \in (s_1 - \varepsilon, s_1 + \varepsilon)$ is equal to the angle between the binormal vectors $\vec{\beta}_1(s)$ and $\vec{\beta}_2(s)$. Thus, $|\vec{\beta}_2(s_2) - \vec{\beta}(s_1)| = 2 \sin \frac{\Delta\theta}{2}$. From this it follows that

$$\lim_{\Delta s \to 0} \frac{\Delta\theta}{\Delta s} = \lim_{\Delta\theta \to 0} \frac{\Delta\theta}{2 \sin \frac{\Delta\theta}{2}} \cdot \lim_{s_2 \to s_1} \frac{|\vec{\beta}(s_2) - \vec{\beta}(s_1)|}{|s_2 - s_1|} = |\vec{\beta}'(s_1)|.$$

It remains to prove the existence of $\vec{\beta}'(s_1)$, where $\vec{\beta}(s) = \frac{\vec{r}' \times \vec{r}''}{|\vec{r}' \times \vec{r}''|}$.

The denominator of this expression for $s = s_1$ is nonzero, since $k(s_1) = \frac{|\vec{r}' \times \vec{r}''|}{|\vec{r}'|^3} \neq 0$. Thus from the conditions of the theorem and the differentiation rule for quotients the existence of $\vec{\beta}'(s_1)$ follows. Transform the formula $\kappa = |\vec{\beta}'(s_1)|$ to a more convenient form for calculations. By definition, $\vec{\beta}(s) = \vec{\tau}(s) \times \vec{\nu}(s)$. Hence $\vec{\beta}' = \vec{\tau}' \times \vec{\nu} + \vec{\tau} \times \vec{\nu}'$. Since $\vec{\tau}' = \vec{r}''_{ss}$, but $\vec{\nu} = \frac{\vec{r}''_{ss}}{|\vec{r}''_{ss}|}$, then $\vec{\tau}' \times \vec{\nu} = 0$ and $\vec{\beta}' = \vec{\tau} \times \vec{\nu}'$. From this it follows that $\vec{\beta}' \perp \vec{\tau}$, and also that $\vec{\beta}' \perp \vec{\beta}$. Hence $\vec{\beta}' = \lambda \vec{\nu}$, where $|\lambda| = |\vec{\beta}'|$. So, $\kappa = |\vec{\beta}'| = |\langle \vec{\beta}', \vec{\nu} \rangle|$, or $\kappa = |\langle \vec{\tau} \times \vec{\nu}', \vec{\nu} \rangle| = |(\vec{\tau} \cdot \vec{\nu} \cdot \vec{\nu}')|$. Find $\vec{\nu}'$:

$$\vec{\nu} = \frac{\vec{r}''_{ss}}{|\vec{r}''_{ss}|}.$$

Hence $\vec{\nu}' = \frac{\vec{r}'''_{sss}}{|\vec{r}''_{ss}|} + A\vec{r}''$, where A is some function of s. Substituting the expression for $\vec{\nu}'$ into the formula for κ, we obtain

$$\kappa = \frac{(\vec{\tau} \cdot \vec{\nu} \cdot \vec{\nu}')}{|\vec{r}''|} = \frac{|(\vec{r}' \cdot \vec{r}'' \cdot \vec{r}''')|}{|\vec{r}''|}.$$

If $\vec{r} = \vec{r}(t)$ is an arbitrary parameterization of class C^3, then

$$\vec{r}'_s = \frac{\vec{r}'_t}{|\vec{r}'_t|}, \quad \vec{r}''_{ss} = \frac{\vec{r}''_{tt}}{|\vec{r}'_t|^2} + A\vec{r}'_t, \quad \vec{r}'''_{sss} = \frac{\vec{r}'''_{ttt}}{|\vec{r}'_t|^3} + B\vec{r}''_{tt} + C\vec{r}'_t, \quad |\vec{r}''_{ss}| = \frac{|\vec{r}'_t \times \vec{r}''_{tt}|}{|\vec{r}'_t|^3},$$

where A, B, and C are some functions of t. Thus

$$\kappa = \frac{|(\vec{r}'_t \cdot \vec{r}''_{tt} \cdot \vec{r}'''_{ttt})|}{|\vec{r}'_t \times \vec{r}''_{tt}|^2}. \qquad \square$$

One may define the sign of the torsion for curves $\gamma \subset \mathbb{R}^3$: as we saw before, $|\vec{\beta}'| = |\langle \vec{\beta}', \vec{\nu} \rangle|$. Define the torsion by the formula $\kappa = -\langle \vec{\beta}', \vec{\nu} \rangle$. Geometrically, this means that the torsion is positive if while moving along a curve the basis $\{\vec{\tau}, \vec{\nu}, \vec{\beta}\}$ turns around $\vec{\tau}$ by the *right-hand rule*; i.e., clockwise as viewed from the initial point of the vector $\vec{\tau}$.

Exercise 1.8.1. Find all curves of nonzero curvature and of zero torsion at all their points.

Solution. Let $\kappa = 0$. Since $|\vec{\beta}'| = \kappa$ for the arc length parameterization, then $\vec{\beta}' = 0$. Consequently, $\vec{\beta}(s) = \vec{\beta}_0$. Moreover, we already know that $\langle \vec{\beta}, \vec{\tau} \rangle = 0$ or $\langle \vec{r}', \vec{\beta}_0 \rangle = 0$, from which it follows that $\langle \vec{r}(s) - \vec{r}(s_0), \vec{\beta}_0 \rangle = \text{const}$. Hence the curve γ is located in a plane orthogonal to the vector $\vec{\beta}_0$; i.e., γ is a plane curve.

Note that the condition $k \neq 0$ is not superfluous. Consider a curve γ, consisting of two arcs γ_1 and γ_2; γ_1 is given by the equations $y = x^4$, $z = 0$ ($0 \leq x < \infty$), and γ_2 is given by the equations $y = 0$, $z = x^4$ ($-\infty < x \leq 0$). The curvature of the obtained curve γ is defined at all points and is zero only at the point $M_0(0, 0, 0)$; the torsion is zero at every point where it is defined. Meanwhile, this γ is not a plane curve. $\qquad \square$

1.8.1 Formulas for Calculations

$$\kappa = \frac{(\vec{r}' \cdot \vec{r}'' \cdot \vec{r}''')}{|\vec{r}''|^2} \quad \text{for the arc length parameterization} \quad \vec{r} = \vec{r}(s),$$

$$\kappa = \frac{(\vec{r}' \cdot \vec{r}'' \cdot \vec{r}''')}{|\vec{r}' \times \vec{r}''|^2} \quad \text{for an arbitrary parameterization} \quad \vec{r} = \vec{r}(t).$$

1.9 The Frenet Formulas and the Natural Equation of a Curve

For all points of a regular curve γ of class C^3, where the curvature is nonzero, three mutually orthogonal unit vectors $\vec{\tau}$, $\vec{v}$, and $\vec{\beta}$ (forming a basis) are uniquely defined. Hence, any vector can be presented as their linear combination. In particular, the vectors $\vec{\tau}'$, $\vec{v}'$, and $\vec{\beta}'$ can be decomposed in terms of $\vec{\tau}$, $\vec{v}$, and $\vec{\beta}$. If a parameterization of a curve is natural, then the coefficients of the decomposition have a geometrical sense, and they are expressed using the curvature and the torsion. Indeed,

$$\vec{\tau}(s) = \vec{r}'(s), \qquad \vec{\tau}' = \vec{r}'' = |\vec{r}''| \cdot \frac{\vec{r}''}{|r''|} = k\vec{v},$$

and $\vec{\beta}' = \lambda \vec{v}$, but from the definition torsion sign it follows that $\lambda = -\kappa$. Hence,

$$\vec{\beta}' = -\kappa \vec{v}.$$

The three vectors $\{\vec{\tau}, \vec{v}, \vec{\beta}\}$ form an orthogonal frame. Hence,

$$\vec{v} = \vec{\beta} \times \vec{\tau}, \qquad \vec{\beta} = \vec{\tau} \times \vec{v}, \qquad \vec{\tau} = \vec{v} \times \vec{\beta},$$

$$\vec{v}' = \vec{\beta}' \times \vec{\tau} + \vec{\beta} \times \vec{\tau}' = -\kappa(\vec{v} \times \vec{\tau}) + k(\vec{\beta} \times \vec{v}) = \kappa\vec{\beta} - k\vec{\tau}.$$

In such a way, we obtain formulas that are called the *Frenet formulas*:

$$\begin{cases} \vec{\tau}' = & k\vec{v}, \\ \vec{v}' = -k\vec{\tau} & +\kappa\vec{\beta}, \\ \vec{\beta}' = & -\kappa\vec{v}, \end{cases} \tag{1.30}$$

or in matrix form,

$$\begin{pmatrix} \vec{\tau}' \\ \vec{v}' \\ \vec{\beta}' \end{pmatrix} = \begin{pmatrix} 0 & k & 0 \\ -k & 0 & \kappa \\ 0 & -\kappa & 0 \end{pmatrix} \begin{pmatrix} \vec{\tau} \\ \vec{v} \\ \vec{\beta} \end{pmatrix}.$$

Using the Frenet formulas, it is easy to find the orthogonal projections of a curve onto the *osculating plane* $(\vec{\tau}, \vec{v})$, onto the *normal plane* $(\vec{v}, \vec{\beta})$, and onto the *rectifying plane* $(\vec{\tau}, \vec{\beta})$. Introduce a Cartesian coordinate system with the origin at a given point $P = \vec{r}(s_0)$ and with the coordinate axes directed along $\vec{\tau}$, $\vec{v}$, and $\vec{\beta}$; i.e., $\vec{\tau} = \vec{i}$, $\vec{v} = \vec{j}$, $\vec{\beta} = \vec{k}$. Apply Taylor's formula to the vector function $\vec{r} = \vec{r}(s)$:

$$\vec{r}(s) = \vec{r}(s_0) + \vec{r}'(s)(s - s_0) + \frac{1}{2}\vec{r}''(s_0)(s - s_0)^2 + \frac{1}{6}\vec{r}'''(s_0)(s - s_0)^3 + \vec{o}((s - s_0)^3).$$

Since

$$\vec{r}(s_0) = 0, \qquad \vec{r}'(s_0) = \vec{\tau}(s_0)\vec{i}, \qquad \vec{r}''(s_0) = k\vec{v} = k\vec{j}, \tag{1.31}$$
$$\vec{r}'''(s_0) = k'\vec{v}(s_0) + k\vec{v}'(s_0) = k'\vec{v}(s_0) + k(-k\vec{\tau}(s_0) + \kappa\vec{\beta}(s_0))$$
$$= -k^2\vec{i} + k\vec{j} + k\kappa\vec{k},$$

then

$$\begin{cases} x(s) = -s_0 - \frac{1}{6}k^2(s - s_0)^3 + \bar{o}_1((s - s_0)^3), \\ y(s) = \frac{1}{2}k(s - s_0)^2 + \frac{1}{6}k'(s - s_0)^3 + \bar{o}_2((s - s_0)^3), \\ z(s) = \frac{1}{6}k\kappa(s - s_0)^3 + \bar{o}_3((s - s_0)^3). \end{cases}$$

From the last formulas we obtain the equations of the orthogonal projections

(1) onto the osculating plane: $y = \frac{1}{2}kx^2 + \bar{o}(x^3)$,
(2) onto the normal plane: $z^2 = Ay^3 + \bar{o}(y^3)$, where $A = \frac{2\kappa^2}{9k}$,
(3) onto the rectifying plane: $z = Bx^3 + \bar{o}(x^3)$, where $B = \frac{1}{6}k\kappa$.

From the deduced equations it follows that the curvature of the projection of a curve onto the osculating plane is equal to the curvature of a given curve at this point, and the curvature of the projection onto the normal plane (at this point) is zero. This explains the notions of these planes.

For space curves there is a theorem analogous to the above theorem for the plane curves.

Theorem 1.9.1 (Fundamental Theorem of Curves). *Let $k(s)$ be an arbitrary continuous positive function, and $\kappa(s)$ an arbitrary continuous function, $0 \leq s \leq a$. Then there is a unique (up to position in space) oriented curve for which $k(s)$ is the curvature, and $\kappa(s)$ is the torsion at the point corresponding to the end of the arc with arc length s.*

Proof. If a curve γ with the given functions of curvature $k(s)$ and torsion $\kappa(s)$ exists, then the Frenet formulas (1.30) are satisfied for it. Hence to deduce the equations of γ it is natural to consider the Frenet formulas as a linear system of ODEs with the given functions $k(s)$ and $\kappa(s)$ as coefficients, in which we solve for the vector functions $\vec{\tau}(s)$, $\vec{v}(s)$, and $\vec{\beta}(s)$. One should also find the vector function $\vec{r}(s)$ using $\vec{\tau}(s)$.

Let the functions $\vec{\tau}(s)$, $\vec{v}(s)$, and $\vec{\beta}(s)$ be the solution of the system (1.30) with the initial conditions $\vec{\tau}(0) = \vec{\tau}_0$, $\vec{v}(0) = \vec{v}_0$, and $\vec{\beta}(0) = \vec{\beta}_0$, and

$$\langle \vec{\tau}_0, \vec{\tau}_0 \rangle = \langle \vec{v}_0, \vec{v}_0 \rangle = \langle \vec{\beta}_0, \vec{\beta}_0 \rangle = 1,$$
$$\langle \vec{\tau}_0, \vec{v}_0 \rangle = \langle \vec{\tau}_0, \vec{\beta}_0 \rangle = \langle \vec{v}_0, \vec{\beta}_0 \rangle = 0, \qquad (1.32)$$
$$(\vec{\tau}_0 \cdot \vec{v}_0 \cdot \vec{\beta}_0) = 1.$$

We wish to prove that the equalities (1.32) hold for any s. Introduce six new functions,

$$\xi_1 = \langle \vec{\tau}(s), \vec{\tau}(s) \rangle, \qquad \xi_2 = \langle \vec{v}(s), \vec{v}(s) \rangle, \qquad \xi_3 = \langle \vec{\beta}(s), \vec{\beta}(s) \rangle,$$
$$\xi_4 = \langle \vec{\tau}(s), \vec{v}(s) \rangle, \qquad \xi_5 = \langle \vec{\tau}(s), \vec{\beta}(s) \rangle, \qquad \xi_6 = \langle \vec{v}(s), \vec{\beta}(s) \rangle,$$

and find the first derivatives of the functions ξ_i using the Frenet formulas:

$$\begin{cases} \xi_1' = \langle \vec{\tau}, \vec{\tau}' \rangle = 2\langle \vec{\tau}', \vec{\tau} \rangle = 2k\xi_4, \\ \xi_2' = 2\langle \vec{v}', \vec{v} \rangle = 2\langle -k\vec{\tau} + \kappa\vec{\beta}, \vec{v} \rangle = -2k\xi_4 + 2\kappa\xi_6, \\ \xi_3' = 2\langle \vec{\beta}', \vec{\beta} \rangle = 2\langle -\kappa\vec{v}, \vec{\beta} \rangle = -2\kappa\xi_6, \\ \xi_4' = \langle \vec{\tau}', \vec{v} \rangle + \langle \vec{\tau}, \vec{v}' \rangle = k\xi_2 - k\xi_1 + \kappa\xi_5, \\ \xi_5' = \langle \vec{\tau}', \vec{\beta} \rangle + \langle \vec{\tau}, \vec{\beta}' \rangle = k\xi_6 - \kappa\xi_4, \\ \xi_6' = \langle \vec{v}', \vec{\beta} \rangle + \langle \vec{v}, \vec{\beta}' \rangle = -\kappa\xi_2 + \kappa\xi_3 - k\xi_5. \end{cases} \qquad (1.33)$$

Consider the above system of equalities as a linear system of ODEs for unknown functions ξ_i, $i = 1, \ldots, 6$, satisfying the initial conditions $\xi_1 = \xi_2 = \xi_3 = 1$, $\xi_4 = \xi_5 = \xi_6 = 0$. From the uniqueness theorem for linear systems of ODEs it follows that

$$\xi_1(s) = \xi_2(s) = \xi_3(s) \equiv 1, \qquad \xi_4(s) = \xi_5(s) = \xi_6(s) \equiv 0, \qquad (1.34)$$

and hence $(\vec{\tau}(s) \cdot \vec{v}(s) \cdot \vec{\beta}(s)) = 1$ by continuity.

Now define the vector function $\vec{r}(s)$ by the formula

$$\vec{r}(s) = \vec{r}_0 + \int_0^s \vec{\tau}(s) \, ds.$$

A curve γ given by this parameterization is required. Indeed, $|\vec{r}'(s)| = |\vec{\tau}(s)| = 1$. Thus s is an arc length parameter counted from the point $\gamma(0) = \vec{r}(0)$. Hence, in view of the Frenet formulas,

$$k(s) = |\vec{r}''(s)| = |\vec{\tau}'(s)| = |k\vec{v}(s)| = k(s).$$

Finally, $\kappa = -\langle \vec{\beta}', \vec{v} \rangle = -\langle -\kappa\vec{v}, \vec{v} \rangle = \kappa(s)$, again in view of the Frenet formulas. Note that γ, generally speaking, is not a curve of class C^3, but only of class C^2. Nevertheless, it has torsion at each point. Thus, we obtain the following result:

Corollary 1.9.1. *If the curvature of a curve γ is continuous, then γ belongs to class C^2, but from the continuity of the torsion it does not follow that γ belongs to class C^3.*

For example, an arbitrary plane curve of class C^2 has zero torsion, but need not belong to class C^3.

Now study the *uniqueness problem* for a curve γ. As we just saw, γ is uniquely defined by the given vectors $\vec{r}_0$, $\vec{\tau}_0$, $\vec{v}_0$, and $\vec{\beta}_0$. Hence, if two curves γ_1 and γ_2 have equal curvature and torsion, as functions of the arc length parameter s, then they differ from each other only by initial conditions $\vec{r}_1(0)$, $\vec{r}_2(0)$ and initial directions of vectors from the triples $\{\vec{\tau}_1(0), \vec{v}_1(0), \vec{\beta}_1(0)\}$ and $\{\vec{\tau}_2(0), \vec{v}_2(0), \vec{\beta}_2(0)\}$. Moving the point $\vec{r}_2(0)$ to the point $\vec{r}_1(0)$ by parallel displacement, we can then match the bases $\{\vec{\tau}_2(0), \vec{v}_2(0), \vec{\beta}_2(0)\}$ with the bases $\{\vec{\tau}_1(0), \vec{v}_1(0), \vec{\beta}_1(0)\}$ by a rotation around this point, and γ_1 and γ_2 would coincide.   □

If one drops in this theorem the requirement $k > 0$, and assumes only that $k(s) \geq 0$ and $k(s) = 0$ at a finite number of points $s_1, s_2, \ldots, s_k$, then it is possible to prove the existence of a curve γ whose curvature coincides with a given function $k(s)$, and the torsion coincides with the function $\kappa(s)$ at all points, expect the points $s_1, s_2, \ldots, s_k$. In fact, the torsion of γ is not defined for these points. The uniqueness of γ does not hold in this case. The curve γ consists of the "rigid" arcs between the points $\gamma(s_i)$ and $\gamma(s_{i+1})$ $(i = 1, \ldots, k - 1)$, but these arcs can be rotated at the points $\gamma(s_i)$ around the vectors $\vec{\tau}(s_i)$.

Example 1.9.1. At the end, derive the curvature and the torsion of a helix $x = a\cos t$, $y = a\sin t$, $z = bt$, where $a > 0$ and b are real numbers. We have

$$
\begin{aligned}
\vec{r} &= \vec{r}(t) = a\cos t\vec{i} + a\sin t\vec{j} + bt\vec{k}, \\
\vec{r}' &= -a\sin t\vec{i} + a\cos t\vec{j} + b\vec{k}, \\
\vec{r}'' &= -a\cos t\vec{i} - a\sin t\vec{j}, \\
\vec{r}''' &= a\sin t\vec{i} - a\cos t\vec{j},
\end{aligned}
\tag{1.35}
$$

and

$$|\vec{r}'| = \sqrt{a^2 + b^2},$$

$$\vec{r}' \times \vec{r}'' = \begin{vmatrix} \vec{i} & \vec{j} & \vec{k} \\ -a\sin t & a\cos t & b \\ -a\cos t & -a\sin t & 0 \end{vmatrix} = ab\sin t\vec{i} - ab\cos t\vec{j} + a^2\vec{k},$$

$$|\vec{r}' \times \vec{r}''| = a\sqrt{a^2 + b^2} \quad \Rightarrow \quad k(t) = \frac{a\sqrt{a^2 + b^2}}{(a^2 + b^2)^{\frac{3}{2}}} = \frac{a}{a^2 + b^2},$$

$$(\vec{r}' \cdot \vec{r}'' \cdot \vec{r}''') = \begin{vmatrix} -a\sin t & a\cos t & b \\ -a\cos t & -a\sin t & 0 \\ a\sin t & -a\cos t & 0 \end{vmatrix} = a^2 b,$$

$$\kappa(t) = \frac{a^2 b}{a^2(a^2 + b^2)} = \frac{b}{a^2 + b^2}.$$

We thus see that the curvature and the torsion of a helix are constants; i.e., do not depend on the parameter t. From Theorem 1.9.1 it follows that *any curve of constant curvature k and constant torsion κ is a helix for which*

$$a = \frac{k}{k^2 + \kappa^2}, \qquad b = \frac{\kappa}{k^2 + \kappa^2}.$$

1.10 Problems: Space Curves

In Section 1.10.1 we consider two problems about curves on a sphere. Then in Section 1.10.2 we shall formulate and solve some problems about space curves.

1.10.1 Spherical Curves

First we give some facts about *spherical geometry*. Let S_R be a sphere of radius R in $\mathbb{R}^3$. If P is some point on S_R, denote by P^* the point diametrically opposite P. If P and Q are arbitrary points on the sphere and $Q \neq P^*$, then there is a unique great circle $C(P, Q)$ containing P and Q (this circle $C(P, Q)$ is the intersection of S_R with the plane through the points P, Q and the center of the sphere S_R). The points P and Q divide $C(P, Q)$ onto two arcs, the smaller one (by arc length) is denoted by PQ. The length of this arc is denoted by $\rho(P, Q)$ or simply by PQ and is called the *distance between the points* P and Q on the sphere S_R.

It turns out that the length of any other curve with endpoints P and Q is greater than PQ. This statement will be proved in Chapter 3, but for now you should accept it without proof. Thus a curve PQ is called a *shortest path*. Assume that the distance between the points P and P^* is equal to πR; we call any arc of the great circle containing P and P^* a *shortest arc* PP^*. It is clear that P and P^* may be joined by a shortest arc not uniquely, but possibly by an infinite number of such arcs. Clearly, a closed curve γ on the sphere S_R divides it into two regions $D_1(\gamma)$ and $D_2(\gamma)$, each of them homeomorphic to a disk.

Definition 1.10.1. A region D on a sphere S_R is *convex* if for any two points P and Q located in D there is the shortest path PQ that belongs to D. A curve γ on a sphere S_R is *convex* if one of the regions $D_1(\gamma)$ or $D_2(\gamma)$ is convex.

We now formulate a sequence of problems.

Problem 1.10.1. Prove that the length of any convex curve γ on the sphere S_R is not greater than $2\pi R$, and that equality holds if and only if γ is a great circle, or a biangle (lune) formed by two shortest paths between some points P and P^*.

Solution. Let D be a convex region bounded by the curve γ.

Consider the case that on γ there exist two diametrically opposite points P, $P^* \in \gamma$. Draw one of the shortest paths $(PP^*)_1$ that is located inside of D. The existence of the shortest path $(PP^*)_1$ follows from the definition of a convex region. The shortest path $(PP^*)_1$ divides D into two regions D_1 and D_2, and the points P and P^* divide the curve γ into two arcs γ_1 and γ_2. Let γ_1 belong to the boundary of D_1, and γ_2 belong to the boundary of D_2, and let $l(\gamma_1) \geq l(\gamma_2)$. Among all shortest paths between P and P^* and containing D_1, take $(PP^*)_2$ that forms a maximal angle with $(PP^*)_1$. (It is not excluded that $(PP^*)_2 = (PP^*)_1$.)

To prove that $\gamma_1 = (PP^*)_2$, assume that $\gamma_1 \neq (PP^*)_2$. Introduce on γ_1 a parameterization $\gamma_1(t)$, where t is the arc length parameter, counting from the point P, $0 \leq t \leq l_1 = l(\gamma_1)$.

Define a function $\alpha(t)$ on $[0, l_1]$ equal to the angle between the shortest paths $P\gamma(t)$ and $(PP^*)_2$ at the point P for $t \neq 0$, and $\alpha(0) = \lim_{t \to 0} \alpha(t)$. The last limit exists in view of the monotonicity of $\alpha(t)$. Analogously define the function $\beta(t)$ as the angle between the shortest paths $P^*\gamma(t)$ and $(P^*P)_2$ at the point P^* for $t < l$ and $\beta(l) = \lim_{t \to l} \beta(t)$. Since $0 = \beta(0) \leq \alpha(0)$ and $\beta(l) > \alpha(l) = 0$, then there is a t_0 such that $\alpha(t_0) = \beta(t_0) \neq 0$. But this means that the curve $P\gamma(t_0) \cup \gamma(t_0)P^*$ is the shortest path joining P with P^* located inside of D and differing from $(PP^*)_2$, which contradicts the definition of $(PP^*)_2$. So, $\gamma_1 = (PP^*)_2$, and hence the length l_1 of the curve γ is πR. But then the inequality

$$\pi R \leq l(\gamma_2) \leq l(\gamma_1) = \pi R$$

holds, from which it follows that γ_2 is the shortest path joining P with P^* and that the length of γ is $2\pi R$.

In the second case draw a polygonal line $p(\gamma)$ inscribed in a curve γ with length different from the length of γ by a sufficiently small value. Denote its vertices by $A_1, A_2, \ldots, A_n$. In view of the convexity of γ, the polygonal line $p(\gamma)$ is also convex, and the inner angle of every vertex in not greater than π:

$$\angle A_i, A_{i+1}A_{i+2} \leq \pi, \qquad i = 1, \ldots, n.$$

Extend the side A_1A_2 to a great circle $C(A_1, A_2)$. Since $p(\gamma)$ does not contain the diametrically opposite points, there is a minimal natural number i_0 such that $A_{i_0} \notin C(A_1, A_2)$.

To prove that $C(A_1, A_2)$ does not intersect the arc γ_1 of a polygonal line $p(\gamma)$ taken from A_{i_0-1} to the point $A_{n+1} = A_1$, assume that $C(A_1, A_2) \cap \gamma_1 \neq \emptyset$. Denote by B_1 the first point of the intersection, counting from A_{i_1-1}, and by B_2 the last point of intersection. Then the circular arcs $A_{i_1-1}B_1$ and B_2A_1 are located outside of $p(\gamma)$. Neither of these arcs is the shortest path, and we obtain contradiction with the convexity property of $p(\gamma)$. So, we have proved that the polygonal line $p(\gamma)$ is located entirely inside a closed semisphere bounded by a circle $C(A_1, A_2)$. From this, by standard methods, one can deduce that the length of $p(\gamma)$ is smaller than the length of $C(A_1, A_2)$, which is equal to $2\pi R$. Hence the length of γ is also smaller than $2\pi R$.   □

Problem 1.10.2. Prove that if γ is a rectifiable closed curve on the sphere S_R and there is a great circle C on the sphere S_R such that the intersection of any closed semicircle C with γ is nonempty, then the length of γ is not smaller than $2\pi R$.

Solution. Let $T = C \cap \gamma$. Take an arbitrary point $A_1 \in T$. The points A_1 and A_1^* divide C into two arcs, C_1 and C_2. Take a point $A_2 \in T$ on the arc C_1 with maximal distance from A_1. In view of the conditions of the problem, such an A_2 exists and differs from A_1. If $A_2 = A_1^*$, then the problem is solved, because the length of each of the arcs γ_1 and γ_2 of γ bounded by the points A_1 and A_1^* is not smaller than πR. If $A_2 \neq A_1^*$, then again in view of the conditions of our problem, there is a point $A_3 \in T$ on the arc $A_1^* D \subset C_2$ of length $\pi R - A_2 A_1^*$.

So, we have obtained three points $A_1, A_2, A_3 \in T$. Each of the arcs $A_1 A_2$, $A_2 A_3$, and $A_3 A_1$ of C is a shortest one, and the sum of their lengths is $2\pi R$. But the same points A_1, A_2, A_3 divide γ into 3 arcs γ_1 (from A_1 to A_2), γ_2 (from A_2 to A_3), and γ_3 (from A_3 to A_1), and their lengths are not smaller than the lengths of certain arcs of C. Hence the length of γ is not smaller than $2\pi R$. □

Problem 1.10.3. Prove that if a simple closed curve divides a sphere S_R into two regions with equal areas, then the length of γ is not smaller than $2\pi R$.

Solution. Denote by D_1 and D_2 two regions into which γ divides S_R. By the condition of the problem, their areas are

$$S(D_1) = S(D_2). \tag{1.36}$$

Let J be the map from a sphere S_R onto itself transforming each point P into a diametrically opposite point P^*; i.e., $J(P) = P^*$. We shall prove that the curves $\gamma^* = J(\gamma)$ and γ have nonempty intersection. Indeed, if $\gamma^* \cap \gamma = \emptyset$, then either $D_1^* = J(D_1) \subset D_2$ or $D_2^* = J(D_2) \subset D_1$, which is impossible in view of equality (1.36). Hence, $\gamma \cap \gamma^* = T \neq \emptyset$. Let a point $P \in T$; then the points P and P^* belong to γ. Indeed, $P \in \gamma$ and $P \in \gamma^*$. Hence, $P^* = J(P) \in J(\gamma^*) = \gamma$. The points P and P^* divide γ into two arcs γ_1 and γ_2, and each of them has length smaller than πR. Thus the length of γ is not smaller than $2\pi R$. □

Problem 1.10.4. Prove that if the length of a simple closed curve γ on the sphere S_R is smaller than $2\pi R$, then there is an open hemisphere S_R' of the sphere S_R that contains γ.

Solution. A curve γ divides the sphere S_R into two regions $D_1(\gamma)$ and $D_2(\gamma)$. One of these regions, in view of Problem 1.10.3, has area smaller than $2\pi R^2$. Let $D_1(\gamma)$ be this region. Denote by $K(P)$ the disk with center at a point P and a radius $\pi 2/R$. Denote by $S(P)$ the area of the intersection of the disks $D_1(\gamma)$ and $K(P_0)$ and let $S_0 = \inf_{P \in S_R} S(P)$. Let $K(P_0)$ be the disk with the property $S(P_0) = S_0$. Let $C(P)$ be a great circle, the boundary of the disk $K(P)$, and let $M = \gamma \cap C(P_0)$. If $M = \emptyset$, then γ sits inside the open disk $K(P_0^*)$. If $M \neq \emptyset$, then $M \in C'(P_0)$, where $C'(P_0)$ is some open semicircle $C(P_0)$; see Problem 1.10.2.

Denote by Q_1 and Q_2 the endpoints of $C'(P_0)$, and let $a = \min \rho(\gamma, C(P_0) \setminus C'(P_0))$. Now take a point P_1 satisfying the following conditions:

(1) P_1 is located in the region bounded by the arcs $Q_1 P_0 \cup P_0 Q_2$ and $C(P_0) \setminus C'(P_0)$.

(2) The disk $K(P_1)$ contains the points Q_1 and Q_2,

(3) $\frac{\pi}{2}R - \frac{a}{4} < \rho(P_1, C(P_0) \setminus C'(P_0)) < \frac{\pi}{4}R + \frac{a}{4}$.

Then the part of the disk $K(P_1)$ that is equal to $K(P_1) \setminus K(P)$ does not include the points of $D_1(\gamma)$, and the part of the disk $K(P_1)$ that is equal to $K(P_0) \setminus K(P_1)$ does not contain the points of $D_1(\gamma)$, in view of the definition of the integer S_0 and of the disk $K(P_0)$, because in the opposite case the area of the intersection of $D_1(\gamma)$ with $K(P_1)$ would be smaller than S_0.

Consequently, $S_0 = 0$ and $D_1(\gamma) \cap K(P_0) = \gamma \cap C(P_0) = \emptyset$. But then the disk $K(P_1)$ has no common points with γ. $\square$

1.10.2 Space Curves

Let $\gamma(t)$ be a regular space curve of class C^k ($k \geq 1$); let t be its arc length parameter, and let $\vec{e}$ be some unit vector. Draw a straight line a through the point $\gamma(0)$ in the direction of $\vec{e}$, and define the function $p(t)$ as the length of the projection of the vector $\overrightarrow{\gamma(0)\gamma(t)}$ onto the line a, by taking into account the sign; i.e., we assume $p(t) = \langle \overrightarrow{\gamma(0)\gamma(t)}, \vec{e} \rangle$.

Problem 1.10.5. Prove that $\dfrac{dp}{dt} = \langle \vec{\tau}, \vec{e} \rangle$.

Solution. If $\vec{r} = \vec{r}(t)$ is a vector function that defines a curve γ, then $\overrightarrow{\gamma(0)\gamma(t)} = \vec{r}(t) - \vec{r}(0)$ and $p(t) = \langle \vec{r}(t) - \vec{r}(0), \vec{e} \rangle$. From these we obtain $\frac{dp}{dt} = \langle \frac{d\vec{r}}{dt}, \vec{e} \rangle = \langle \vec{\tau}, \vec{e} \rangle$. $\square$

Corollary 1.10.1. *If γ is closed curve with the length l, then*

$$\int_0^l \langle \vec{\tau}, \vec{e} \rangle \, dt = 0 \text{ for any } \vec{e}, \quad |\vec{e}| = 1.$$

Corollary 1.10.2. *Let d be the distance between the endpoints of a curve γ of length l. Then*

$$d = |\overrightarrow{\gamma(0)\gamma(l)}| = \int_0^l \langle \vec{\tau}, \vec{e} \rangle \, dt, \text{ where } \vec{e} = \frac{\overrightarrow{\gamma(0)\gamma(l)}}{|\overrightarrow{\gamma(0)\gamma(l)}|}.$$

Theorem 1.10.1 (Fenchel's problem). *The integral curvature of an arbitrary closed space curve γ is not smaller than 2π; equality holds if and only if γ is a convex plane curve.*

Proof. Let $\gamma(t)$ be an arc length parameterization of a curve γ with t ($0 \leq t \leq l = l(\gamma)$). Define a curve $\sigma(t)$ on a sphere S_1 by associating with each point $\gamma(t)$ the end of the vector $\vec{\tau}(t) = \dot{\gamma}(t)$ whose origin coincides with the center of the sphere S_1. The curve $\gamma(t)$ is called the *indicatrix of a tangent line* to γ. The *integral curvature* of γ, defined by $\int_0^l k(t) \, dt$, is equal to the length l_1 of the indicatrix of the tangent line $\sigma(t)$. Indeed,

$$l_1 = l(\sigma(t)) = \int_0^l |\vec{\tau}'| \, dt = \int_0^l |k| \, dt = \int_0^l k(t) \, dt$$

in view of the Frenet formulas. So, we must show that the length l_1 of the indicatrix $\sigma(t)$ is not smaller than 2π. Assume that $l_1 < 2\pi$. Then, as was proven in Problem 1.10.4, there is an open hemisphere S_1' of the sphere S_1 containing σ. Let P be the center of this hemisphere, and let $\vec{e}$ be the unit vector with the origin at the center of S_1 and with endpoint at P. Then $\int_0^l \langle \vec{\tau}(t), \vec{e} \rangle \, dt > 0$ holds, since $\langle \vec{\tau}, \vec{e} \rangle > 0$ for all t, but on the other hand, from the corollary of Problem 1.10.5 it follows that $\int_0^l \langle \vec{\tau}(t), \vec{e} \rangle \, dt = 0$. This contradiction proves that $l_1 = l(\sigma(t)) = \int_0^l k(t) \, dt$ is not smaller than 2π, and equality holds if and only if $\sigma(t)$ is a great circle on S_R and consequently, γ is a convex plane curve.   □

Consider two curves $\gamma(t)$ and $\tilde{\gamma}(t)$, of which the first one is an arc of a convex plane curve, and the second one is an arbitrary space curve. Let $\gamma(t)$ have endpoints A and B, and $\tilde{\gamma}(t)$ the endpoints $\tilde{A}$ and $\tilde{B}$. Also assume that t is an arc length parameter on both curves, counting from the point A and the point $\tilde{A}$, respectively. Solve the following problem; see [Bl].

Problem 1.10.6 (The twist of a curve). If the lengths of the curves $\gamma(t)$ and $\tilde{\gamma}(t)$ are equal and their curvatures $k(t)$ and $\tilde{k}(t)$ satisfy the inequality $\tilde{k}(t) \le k(t)$, then $AB \le \tilde{A}\tilde{B}$, and equality holds if and only if γ and $\tilde{\gamma}$ are identified by a rigid motion of space. If $\tilde{k}(t) = k(t)$, then this problem can be formulated thus: *twisting a curve increases the distance between its endpoints.* If $\tilde{\gamma}(t)$ is also a plane curve, then in this case our problem is equivalent to Problem 1.7.9 about a bent bow.

Solution. Take a point $\gamma(t_0)$ on γ at which the tangent line is parallel to AB. Place a curve $\tilde{\gamma}$ such that the point $\tilde{\gamma}(t_0)$ coincides with $\gamma(t_0)$, and $\tilde{\vec{\tau}}(t_0) = \vec{\tau}(t_0)$. Denote by $\sigma(t)$ and $\tilde{\sigma}(t)$ the indicatrices of the tangent line of the curves $\gamma(t)$ and $\tilde{\gamma}(t)$, respectively. Then first, the length $\tilde{m}(t)$ of the arc $\tilde{\sigma}(t_0)\tilde{\sigma}(t)$ of the curve $\tilde{\sigma}$ is not greater than the length $m(t)$ of the arc $\sigma(t_0)\sigma(t)$ of the curve σ. Indeed, from the Frenet formulas we obtain

$$\tilde{\vec{\tau}}' = \tilde{k}(t)\tilde{\vec{v}}(t), \quad \vec{\tau}' = k(t)\vec{v}(t), \quad |\tilde{\vec{\tau}}'| = |\tilde{k}(t)| \le k(t) = |\vec{\tau}'|.$$

Thus

$$\tilde{m}(t) = \int_{t_0}^t |\tilde{\vec{\tau}}'(t)| \, dt \le \int_{t_0}^t |\vec{\tau}'| \, dt = m(t). \tag{1.37}$$

Second, if we denote by $\alpha(t)$ and $\tilde{\alpha}(t)$ the angles $\angle\sigma(t_0)O\sigma(t)$ and $\angle\tilde{\sigma}(t_0)O\tilde{\sigma}(t)$, respectively, then the functions $\alpha(t)$ and $\tilde{\alpha}(t)$ should satisfy the inequality

$$\tilde{\alpha}(t) \le \alpha(t). \tag{1.38}$$

Indeed, the distance between the points $\sigma(t_0)$ and $\sigma(t)$ on the sphere S_1 is equal to $m(t)$, and the distance between the points $\tilde{\sigma}(t_0)$ and $\tilde{\sigma}(t)$ on the same sphere

S_1 is not greater than $\tilde{m}(t)$, and moreover, in view of (1.37), it is not greater than $m(t)$. Thus, the central angle $\angle \tilde{\sigma}(t_0) O \tilde{\sigma}(t)$ is not greater than $\angle \sigma(t_0) O \sigma(t)$, and hence the inequality (1.38) is proved.

Note also that by the choice of the point $\gamma(t_0)$, the angle $\alpha(t)$ for $t \leq t_0$ and for $t > t_0$ varies between the values 0 and π. Hence, from the inequality (1.37) we have

$$\cos \tilde{\alpha}(t) = \langle \tilde{\vec{\tau}}'(t), \tilde{\vec{\tau}}(t_0) \rangle \geq \cos \alpha(t) = \langle \vec{\tau}(t), \vec{\tau}(t_0) \rangle. \qquad (1.39)$$

Finally, as we already know (see Problem 1.10.5),

$$AB = \int_0^{t_0} \cos \alpha(t)\, dt + \int_{t_0}^l \cos \alpha(t)\, dt,$$

and the projection $\overline{\tilde{A}\tilde{B}}$ of the line segment $\tilde{A}\tilde{B}$ onto the direction $\vec{\tau}(t_0)$ is

$$\overline{\tilde{A}\tilde{B}} = \int_0^{t_0} \cos \tilde{\alpha}(t)\, dt + \int_{t_0}^l \cos \tilde{\alpha}(t)\, dt.$$

Hence, from the inequality (1.39) we obtain

$$\tilde{A}\tilde{B} \geq \overline{\tilde{A}\tilde{B}} \geq AB,$$

and, as is seen from the text of the solution of the problem, equality in the last inequalities holds if and only if the curves $\sigma(t)$ and $\tilde{\sigma}(t)$ coincide. But then $\gamma(t)$ and $\tilde{\gamma}(t)$ also coincide. □

1.11 Phase Length of a Curve and the Fenchel–Reshetnyak Inequality

It is interesting to understand the restrictions on the integral curvature of a non-closed curve. Let a curve L join two different points A and B. Measure the angles α and β formed by the chord AB with the tangent rays to our curve at its endpoints. If one closes the curve by adding a straight line segment almost parallel to the chord, then the tangent vectors in neighborhoods of A and B are rotated by the angles $\pi - \alpha$ and $\pi - \beta$, respectively. It is easy to ensure that practically the whole integral curvature of a closed arc of a curve will be reduced to these turnings, i.e., that it will be equal to $2\pi - \alpha - \beta$ with arbitrary accuracy. Applying Fenchel's inequality, we now obtain an answer to the above question:

$$\int_L k(s)\, ds \geq \alpha + \beta.$$

This inequality with the above elegant argument is due to Y.G. Reshetnyak. This also contains the previous statement, since one may apply it to both parts of a closed curve supported by a common chord. Thus, each of the above inequalities is easily deduced from the other. The direct proof of either is much more complicated.

Below we give a new proof of these inequalities suggested by V.V. Ivanov. It is based on the notion of a *phase distance between two vectors attached to different points of the space*. The idea of the construction given below is actually contained in the second of the inequalities under discussion. Indeed, the sum $\alpha + \beta$ is defined only by the two endpoints of the curve and by the two directions at which the curve starts and finishes its movement; furthermore, the sum is not only a lower bound, but also an exact lower bound for the integral curvatures of all curves satisfying the same boundary conditions as the curve L; one may check this by repeating the above arguments. From here it is clear that this sum, as a function of a pair related vectors, must satisfy the triangle inequality, and thus the original notion of the length of a smooth curve is naturally related to such a function; this is its phase length. As we shall see, Reshetnyak's inequality means that the phase length of a curve is not smaller than the phase distance between its endpoints. But initially, we prove that the integral curvature and the phase length are the same. This is a matter of analysis.

1. The following discussion holds in Euclidean space of arbitrary dimension.

Definition 1.11.1. The *phase distance from a vector $\vec{a}$ at a point A to a vector $\vec{b}$ at a point B* is the sum of the angles that the vector $\overrightarrow{AB}$ forms with the directions of $\vec{a}$ and $\vec{b}$.

Indeed, this definition makes sense only in the case that $\vec{a}$ and $\vec{b}$ are both nonzero, and the points A and B are different. The properties of the phase distance are a little unusual, though if one takes into account its geometrical sense, they become quite natural.

Lemma 1.11.1. First, the values of the phase distance are bounded:

$$0 \le \varphi(\vec{a}, \vec{b}) \le 2\pi.$$

Second, instead of symmetry, the following identity holds:

$$\varphi(\vec{a}, \vec{b}) + \varphi(\vec{b}, \vec{a}) = 2\pi.$$

And third, the triangle inequality can be written in the following form:

$$\varphi(\vec{a}, \vec{b}) + \varphi(\vec{b}, \vec{c}) + \varphi(\vec{c}, \vec{a}) \ge 2\pi.$$

Proof. Only the last statement requires a short discussion. The configuration consisting of three related vectors, generally speaking, is located in a five-dimensional space, but the inequality corresponding to it reflects elementary properties of a usual trihedral angle in ordinary three-dimensional space. For the proof it is enough to consider the nondegenerate case, in which the points A, B, C at which vectors $\vec{a}$, $\vec{b}$, $\vec{c}$ are attached do not lie on a straight line. Mark on the extensions of the edges CA, AB, and BC of the triangle the points A', B', and C'. The least value of the sum of the angles formed by the vector $\vec{a}$ with the rays AA' and AB is equal to the value of $\angle A'AB$ and is attained only in the case that $\vec{a}$ lies in the plane of $\triangle ABC$ and varies within its specified exterior angle. In the same way, $\angle B'BC$ and $\angle C'CA$ bound from below the sums of the other angles with vertices at B and C. From here it follows that the sum of the three phase distances under discussion, which breaks up into the sum of six angles adjoining the vectors $\vec{a}$, $\vec{b}$, $\vec{c}$, is not smaller than the sum of the exterior angles of $\triangle ABC$, which is equal to double the sum of its interior angles.   □

Using the second statement of the lemma, the triangle inequality for phase distances can be written down by the usual method:

$$\varphi(\vec{a}, \vec{c}) \le \varphi(\vec{a}, \vec{b}) + \varphi(\vec{b}, \vec{c}).$$

Moreover as we have seen, this inequality turns into an exact equality only in the case that the vectors $\vec{a}$, $\vec{b}$, $\vec{c}$ lie in the corresponding exterior angles of $\triangle ABC$.

2.   Consider now a smooth curve L, determined by the arc length parameterization $\vec{r} = \vec{r}(s)$ with $0 \le s \le l$. Let $s_1 < s_2 < \cdots < s_n$, where $s_1 = 0$, $s_n = l$ and to the consecutive values of the parameter there correspond different points on L. Draw at each of these points $\vec{r}(s_i)$ the tangent vector $\vec{\tau}_i = \vec{r}'(s_i)$. The chain thus obtained $\vec{\tau}_1, \vec{\tau}_2, \ldots, \vec{\tau}_n$, is said to be a *phase polygonal line inscribed in the curve* L, and its *phase length* is the sum

$$\varphi(\vec{\tau}_1, \vec{\tau}_2) + \cdots + \varphi(\vec{\tau}_{n-1}, \vec{\tau}_n).$$

The *phase length of the curve* L should be defined as the infimum $\Phi(L)$ of the phase lengths of phase polygonal lines inscribed in it.

Lemma 1.11.2. If a chord joining the endpoints of a nonclosed curve L forms angles α and β with the tangent rays to the curve at its endpoints, then the phase length of any phase polygonal line inscribed in L is not smaller than the sum of the angles α and β. In particular,

$$\Phi(L) \ge \alpha + \beta.$$

If the curve L is closed, then the phase length at any phase polygonal line inscribed in it is not smaller than 2π, so

$$\Phi(L) \ge 2\pi.$$

Proof. Indeed, in the first case, for every phase polygonal line $\vec{\tau}_1, \vec{\tau}_2, \ldots, \vec{\tau}_n$ inscribed in L, the phase distance from $\vec{\tau}_1$ up to $\vec{\tau}_n$ is obviously equal to the sum of the specified angles, so by the triangle inequality,

$$\alpha + \beta = \varphi(\vec{\tau}_1, \vec{\tau}_n) \le \varphi(\vec{\tau}_1, \vec{\tau}_2) + \cdots + \varphi(\vec{\tau}_{n-1}, \vec{\tau}_n).$$

In the case of a closed curve, when $\vec{\tau}_1 = \vec{\tau}_n$, it is enough to apply the triangle inequality to the pair of vectors $\vec{\tau}_1$, $\vec{\tau}_{n-1}$ and to note that $\varphi(\vec{\tau}_1, \vec{\tau}_{n-1}) = 2\pi - \varphi(\vec{\tau}_{n-1}, \vec{\tau}_n)$. $\Box$

By the way, if we take into account Section 1.9, it is easy to show that the inequalities in Lemma 1.11.2 become exact if and only if the curve either is a line segment or lies in some plane where the vertices of any polygonal line inscribed in it serve as consecutive vertices of a convex polygon. This means that the curve is convex as a whole.

3. For studying the relations between the integral curvature of a curve and its phase length, we need some asymptotic formulas describing the local geometry of a twice differentiable curve from the point of view of its curvature.

To simplify writing the expressions, use the symbol $o(\sigma^m)$ as the general designation of any function possessing the property that after division by σ^m it approaches zero as $\sigma \to 0$, uniformly with respect to other variables on which it may depend.

For the arc length parameterization $\vec{r} = \vec{r}(s)$ of a three-times differentiable curve L, the tangent vector $\vec{\tau}(s) = \vec{r}'(s)$ has unit length and is orthogonal to $\vec{\tau}'(s)$, whose length, as was noted, is taken as a measure of the curvature $k(s)$ of the curve L at the point with coordinate s.

Lemma 1.11.3. The following formulas hold:

$$\langle \vec{r}(s+\sigma) - \vec{r}(s), \vec{\tau}(s) \rangle = \sigma - \frac{k^2}{6}\sigma^3 + o(\sigma^3),$$

$$|\vec{h}(s)| = \sigma - \frac{k^2}{4}\sigma^3 + o(\sigma^3),$$

$$\langle \vec{\tau}(s), \vec{\tau}(s+\sigma) \rangle = 1 - \frac{k^2(s)}{2}\sigma^2 + o(\sigma^2),$$

where $\vec{h}(s) = \vec{r}(s+\sigma) - \vec{r}(s)$ and $\sigma > 0$.

Proof. Applying Taylor's formula and the Frenet formulas, we have

$$\vec{r}(s+\sigma) - \vec{r}(s) = \vec{\tau}(s)\sigma + \frac{1}{2}\vec{\tau}'(s)\sigma^2 + \frac{1}{6}\vec{\tau}''(s)\sigma^3 + o(\sigma^4)$$

$$= \vec{\tau}(s)\sigma + \frac{1}{2}k\vec{v}(s)\sigma^2 + \frac{1}{6}[k'\vec{v}(s) - k^2\vec{\tau}(s) + \kappa\vec{\beta}(s)]\sigma^3 + o(\sigma^4).$$

From this and from the properties of the Frenet frame the first formula of the lemma follows. Assume that $\vec{h}(s) = \vec{r}(s+\sigma) - \vec{r}(s)$. Applying the above decomposition for the vector $\vec{h}(s)$, we have

$$|\vec{h}(s)|^2 = \sigma^2 + \frac{1}{4}k^2\sigma^4 - \frac{1}{3}k^2\sigma^4 + o(\sigma^4) = \sigma^2 - \frac{1}{12}k^2\sigma^4 + o(\sigma^4).$$

Taking the square root, we obtain the second equality of the lemma. The last equality of the lemma can be proved analogously. $\Box$

Remark 1.11.1. By elaborating the proof of the lemma, it is possible to obtain a similar statement for a twice continuously differentiable curve.

4. Continuing considerations of a three-times differentiable curve, we return to the question in which we are interested. First of all, write down the Taylor decomposition of the cosine,

$$\langle \vec{\tau}(s), \vec{\tau}(s+\sigma) \rangle = \cos \omega = 1 - \frac{\omega^2}{2} + o(\omega^2),$$

where ω is the *angle between the vectors* $\vec{\tau}(s)$ and $\vec{\tau}(s+\sigma)$, and compare it with the decomposition specified in Lemma 1.11.3. Indeed, we shall come to the well-known equality

$$\omega = k(s)\sigma + o(\sigma),$$

characterizing the curvature as the velocity at which the angle ω grows as the parameter σ varies. As we shall now see, the phase distance, which generally differs from the usual angle between the vectors, is calculated for small parts of a curve by the same formula.

Lemma 1.11.4. The phase distance φ from a vector $\vec{\tau}(s)$ up to a vector $\vec{\tau}(s+\sigma)$ for small $\sigma > 0$ is almost proportional to the arc σ

$$\varphi = k(s)\sigma + o(\sigma).$$

Proof. By definition, φ is a sum of two angles φ_1 and φ_2 formed by the chord $\vec{h} = \vec{r}(s+\sigma) - \vec{r}(\sigma)$ with tangent vectors $\vec{\tau}(s)$ and $\vec{\tau}(s+\sigma)$. For calculating the cosine of the first of these angles, divide the decompositions in Lemma 1.11.3:

$$\cos \varphi_1 = \frac{\langle \vec{h}, \vec{\tau}(s) \rangle}{|\vec{h}|} = 1 - \frac{k^2(s)}{8}\sigma^2 + o(\sigma^2).$$

We have already seen how from here it is possible to obtain the asymptotic formula

$$\varphi_1 = \frac{k(s)}{2}\sigma + o(\sigma).$$

By reversing the direction of the curve (which was determined by the choice of parameterization, but played a minor role) the previous equality will transform to a similar expression for the cosine of the second angle φ_2. One needs only to replace $k(s)$ by $k(s+\sigma)$. However, it is possible to work without this replacement, as the difference between the two values as a function of σ is infinitesimally small uniformly with respect to s. Thus,

$$\varphi_2 = \frac{k(s)}{2}\sigma + o(\sigma),$$

and it remains only to combine the two formulas into one.   □

The following theorem, an integral version of Lemma 1.11.4, summarizes the discussion so far. Comparing it with Lemma 1.11.2 allows us to look at the Fenchel–Reshetnyak inequality from a new point of view.

Theorem 1.11.1. *The phase length of a twice differentiable curve is equal to its integral curvature,*

$$\Phi(L) = \int_L k(s)\, ds.$$

Proof. Select $s_1 < s_2 < \cdots < s_n$, where $s_1 = 0$ and $s_n = l$, in a closed interval $0 \le s \le l$. If all differences $\Delta s_i = s_{i+1} - s_i$ are small enough, then the neighboring points $\vec{r}(s_i)$ and $\vec{r}(s_{i+1})$ cannot coincide, and we have an opportunity to compose the sum

$$S = \sum_{i=1}^{n-1} \varphi(\vec{\tau}(s_i), \vec{\tau}(s_{i+1})).$$

The addition of new vertices to a phase polygonal line, in view of the triangle inequality, does not decrease its phase length. Therefore, the phase length of the initial curve is equal to the limit of the sums under discussion, provided that the greatest of the arcs Δs_i tends to zero. According to Lemma 1.11.4, the sum S is expressed as

$$S = \sum_{i=1}^{n-1} k(s_i)\Delta s_i + \sum_{i=1}^{n-1} \varepsilon_i \Delta s_i,$$

where all ε_i uniformly tend to zero and simultaneously with Δs_i. Thus, the second term of the last expression vanishes in the limit, and the first, as the integral sum for the curvature, approaches its integral. $\square$

1.12 Exercises to Chapter 1

Exercise 1.12.1. Let O be a point on a circle of radius a, and a ray intersecting the circle at a varying point A rotates around O. Mark off two line segments $AM_1 = AM_2 = 2a$ on this ray on two sides of A. Deduce the equation of a curve traced by the points M_1 and M_2.
Answer: a cardioid.

Exercise 1.12.2. A disk of radius a rolls along a straight line without sliding. Deduce the parametric equations of the curve traced by the point M on the boundary circle.
Answer: a cycloid.

Exercise 1.12.3. What curve is plotted by the parametric equations $x = a \log(t)$, $y = \frac{a}{2}(t + \frac{1}{t})$?

Exercise 1.12.4. Find the projections of the curve $x = t$, $y = t^2$, $z = t^3$ onto the coordinate planes.

Exercise 1.12.5. Prove that the projection onto the plane YOZ of the intersection curve of an *elliptical paraboloid* $x = y^2 + z^2$ with a plane $x - 2y + 4z = 0$ is a circle of radius $R = 3$ centered at $M(0, 1, -2)$.

Exercise 1.12.6. Find conditions for the existence of an asymptote to the space curve $x = x(t)$, $y = y(t)$, $z = z(t)$ as t goes to infinity.

Exercise 1.12.7. Deduce the equations of the normal line and the tangent plane to the helix $x = 2\cos t$, $y = 2\sin t$, $z = 4t$ at the point $t = 0$.

Exercise 1.12.8. Deduce the equation of the tangent line to the curve $x = \cosh t$, $y = \sinh t$, $z = ct$.

Exercise 1.12.9. Deduce the equation of the main normal and binormal to the curve

$$x = t, \quad y = t^2, \quad z = t^3, \quad t = 1.$$

Exercise 1.12.10. Deduce the equation of the main normal and binormal to the curve $xy = z^2$, $x^2 + y^2 = z^2 + 1$ at the point $M(1, 1, 1)$.

Exercise 1.12.11. Deduce the equation of an osculating plane to the intersection curve of the sphere $x^2 + y^2 + z^2 = 9$ and the hyperbolic cylinder $x^2 - y^2 = 3$ at the point $M(2, 1, 2)$.

Exercise 1.12.12. Deduce the equation of the tangent line to the curve defined by the equations $x^2 + y^2 + z^2 = 1$, $x^2 + y^2 = x$ at the point $M(0, 0, 1)$.

Exercise 1.12.13. Mark off the line segments of fixed length on the binormals to a simple helix. Find the equation of a curve that is traced by the endpoints of these line segments.

Exercise 1.12.14. Prove the following generalizations of the well-known Lagrange *intermediate value theorem* for the case of a smooth space curve.

1. Let A and B be the endpoints of a smooth space curve $\gamma(t)$, and let Π be a smooth family of tangent planes to γ. Then there is a value of the parameter t_0 such that the plane $\Pi(t_0)$ is parallel to the line segment AB.
2. Let A and B be the endpoints of a smooth space curve $\gamma(t)$ that belongs to a closed convex surface. Then there is a value of the parameter t_0 such that the tangent plane at the point $\gamma(t_0)$ is parallel to the line segment AB.

Exercise 1.12.15. Find the curvature and the torsion of the following curves:

(a) $x = \exp(t)$, $y = \exp(-t)$, $z = t\sqrt{2}$,

(b) $x = \cos^3 t$, $y = \sin^3 t$, $z = \cos(2t)$.

Exercise 1.12.16. At what points of the curve $x = a(t - \sin t)$, $y = a(1 - \cos t)$, $z = 4a\cos\frac{t}{2}$ does the radius of curvature take its local maximum?

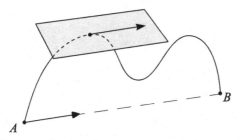

Figure 1.27. Generalization of Lagrange's intermediate value theorem.

Exercise 1.12.17. Prove that

$$(1) \qquad (\vec{\tau} \cdot \vec{\beta} \cdot \vec{\beta}') = \kappa,$$

$$(2) \qquad (\vec{\beta}' \cdot \vec{\beta}'' \cdot \vec{\beta}''') = \kappa^2 \left(\frac{k}{\kappa}\right)',$$

$$(3) \qquad (\vec{\tau}' \cdot \vec{\tau}'' \cdot \vec{\tau}''') = k^3 \left(\frac{\kappa}{k}\right)'.$$

Exercise 1.12.18. Find the length of the curve $x = a \cosh t$, $y = a \sinh t$, $z = at$ between the points 0 and t.

Exercise 1.12.19. Find the length of the *astroid* $x = a \cos^3 t$, $y = a \sin^3 t$.

Exercise 1.12.20. Find the length of the *cycloid* $x = a(t - \sin t)$, $y = a(1 - \cos t)$ $(0 \le t \le 2\pi)$.

Exercise 1.12.21. Find the curvature of the curve defined by the equations in implicit form

$$x + \sinh x = y + \sinh y, \qquad z + \exp z = x + \log(1 + x) + 1$$

at the point $M(0, 0, 0)$.

Exercise 1.12.22. Find an evolute of the *tractrix* $x = -a\left(\log \tan \frac{t}{2} + \cos t\right)$, $y = a \sin t$.

Exercise 1.12.23. Find evolvent of the circle $x^2 + y^2 = R^2$.

Exercise 1.12.24. Prove the constancy of the ratio of curvature to the torsion of the curve

$$x = a \int_{t_0}^{t} \sin \alpha(t) \, dt, \quad y = a \int_{t_0}^{t} \cos \alpha(t) \, dt, \quad z = bt.$$

Exercise 1.12.25.* Prove that a smooth curve $\gamma(s)$ lies on a unit sphere if and only if the following equality holds:

$$(k')^2 = k^2 \kappa^2 (k^2 - 1)$$

and $k \ge 1$, where $k = k(s)$, $\kappa = \kappa(s)$ are the curvature and the torsion of the curve $\gamma(s)$.

2
Extrinsic Geometry of Surfaces in Three-dimensional Euclidean Space

2.1 Definition and Methods of Generating Surfaces

Definition 2.1.1. A connected set Φ in $\mathbb{R}^3$ is said to be a *two-dimensional surface* if for an arbitrary point $P \in \Phi$ there exist an open ball U_P in $\mathbb{R}^3$ with center at P and a continuous injective map $\psi : U_P \to \mathbb{R}^3$ such that ψ maps $W = \Phi \cap U_P$ onto an open disk D_1 of radius 1 on some plane α in the space $\mathbb{R}^3$.

In this definition, a "little" disk D_1 can be replaced by an arbitrary open set[3] of a plane α, diffeomorphic to a disk. See Section 2.6 with further discussion of the notion of a surface.

Introduce in the plane α Cartesian orthogonal coordinates u, v with the origin at the center of the disk D_1. Denote by φ_P the inverse of the restriction of the map ψ to D_1. Then $\varphi_P(D_1) = W$, and the map $\varphi_P : D_1 \to \mathbb{R}^3$ defines a vector function $\vec{r} = \vec{r}(u, v) = x(u, v)\vec{i} + y(u, v)\vec{j} + z(u, v)\vec{k}$, where $u^2 + v^2 < 1$. We thereby obtain that a surface Φ in some neighborhood of one of its (arbitrary) points can be determined by three functions of two variables:

$$x = x(u, v), \quad y = y(u, v), \quad z = z(u, v).$$

These functions are called a *parameterization* of a surface. A parameterization of a surface Φ is said to be *k-fold continuously differentiable* if the functions $x(u, v)$, $y(u, v)$, and $z(u, v)$ are actually k-fold continuously differentiable. The

---

[3] A subset $U \subset \mathbb{R}^n$ is *open* if for every point $x \in U$ there is a number $\varepsilon > 0$ such that $y \in U$ whenever $\|x - y\| < \varepsilon$.

set $W = \varphi_P(D_1)$ is called a *coordinate neighborhood* of the point P on the surface Φ.

Definition 2.1.2. A surface Φ is *k-fold continuously differentiable* if in some neighborhood of each of its points there is a k-fold continuously differentiable parameterization $x = x(u, v)$, $y = y(u, v)$, $z = z(u, v)$. In this case, we say that the surface Φ is of class C^k.

Definition 2.1.3. A surface Φ of class C^k ($k \geq 1$) is *regular* if for each of its points P there is a parameterization $\varphi_P : D_1 \rightarrow \mathbb{R}^3$ of class C^k with maximal rank.[4]

The last condition can be written in the form $\vec{r}_u \times \vec{r}_v \neq 0$, where

$$\vec{r}_u = x_u(u, v)\vec{i} + y_u(u, v)\vec{j} + z_u(u, v)\vec{k}, \quad \vec{r}_v = x_v(u, v)\vec{i} + y_v(u, v)\vec{j} + z_v(u, v)\vec{k}.$$

There are other presentations of a surface besides the parametric ones.

Explicitly given surface. Let $f : D_1 \subset \mathbb{R}^2 \rightarrow \mathbb{R}^1$ be a function of class C^k ($k \geq 1$). Then the set of points

$$\{(x, y, f(x, y)) : (x, y) \in D_1\},$$

the graph of a function $f(x, y)$, forms a regular surface of class C^k. The equation of the surface in this case is usually written as $x = x$, $y = y$, $z = f(x, y)$ or simply as $z = f(x, y)$, $(x, y) \in D_1$.

Implicitly given surface. Let D be some open connected set of the space $\mathbb{R}^3$, let $H : D \rightarrow \mathbb{R}^1$ be a differentiable map of class C^k, and let zero be a regular value of the map H. Then each connected component of the set $\Phi = H^{-1}(0)$ is a regular surface of class C^k. The equation of the surface in this case can be written in the form $H(x, y, z) = 0$; it is called an *implicit equation of the surface* Φ. If a regular surface Φ is given by *parametric equations*

$$x = x(u, v), \quad y = y(u, v), \quad z = z(u, v),$$

then for each point $P \in \Phi$ there is a neighborhood in which Φ can be presented by an explicit equation. Indeed, since the rank of the map φ_P at the point P is 2, then by the inverse function theorem the variables u and v can be expressed through x, y; through x, z; or through y, z; and then the equation of Φ in a neighborhood of P can be written in one of the following forms:

$$z = z(u(x, y), v(x, y)) = f_1(x, y),$$
$$y = y(u(x, z), v(x, z)) = f_2(x, z),$$
$$x = x(u(y, z), v(y, z)) = f_3(y, z).$$

---

[4] A C^k-*coordinate patch* for some $k \geq 1$ is a one-to-one C^k-map $\mathbf{x} : D_1 \rightarrow \mathbb{R}^3$ with maximal rank.

But if Φ is defined by the implicit equation $H(x, y, z) = 0$, then since zero is a regular value of the map H, we obtain that one of derivatives H_x, H_y, H_z differs from zero at every point $P \in \Phi$. If, for instance, $H_z \neq 0$, then by the implicit function theorem there is a function $f(x, y)$ such that $H(x, y, f(x, y)) \equiv 0$, and consequently, a surface Φ in some neighborhood of the point P can be defined by the explicit equation $z = f(x, y)$.

Example 2.1.1. The one-sheeted cone given by $z = \sqrt{x^2 + y^2}$ is not a regular surface; $(0, 0, 0)$ is its singular point. The ellipsoid $\frac{x^2}{a^2} + \frac{y^2}{b^2} + \frac{z^2}{c^2} = 1$ is a regular surface. This follows from the fact that 0 is a regular value of the function $H = \frac{x^2}{a^2} + \frac{y^2}{b^2} + \frac{z^2}{c^2} - 1$.

Definition 2.1.4 (Local coordinate system). Let $\vec{r} = \vec{r}(u, v)$ be parametric equations of a surface Φ in some neighborhood of a point $P \in \Phi$. Then to each point of this neighborhood corresponds an ordered pair of real numbers (u, v), which are said to be *local coordinates of the point*. An arbitrary curve γ on Φ can be defined by equations in local coordinates $u = u(t), v = v(t)$ $(a \le t \le b)$. The equations of γ in the space $\mathbb{R}^3$ take the form

$$x = x(u(t), v(t)) = \tilde{x}(t), \quad y = y(u(t), v(t)) = \tilde{y}(t), \quad z = z(u(t), v(t)) = \tilde{z}(t).$$

The curves $u = t, v = $ const and $u = $ const, $v = t$ are called the *coordinate curves*, namely, *u-curves* and *v-curves*.

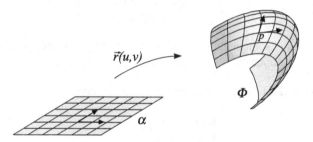

Figure 2.1. Local coordinate system.

2.1.1 Special Coordinate Systems

Let $\varphi_1 \colon D_1 \to \Phi$ be some regular parameterization of a surface Φ of class C^k $(k \ge 1)$, and let $\mathbf{h} \colon D_P \to D_1$ be an injective map of a disk D_P in D_1 of class C^k $(k \ge 1)$ with nonzero Jacobian $\det J$. Then composition of the maps $\varphi_P = \varphi_1 \circ \mathbf{h} \colon D_P \to \Phi$ is also a regular parameterization of Φ of the same class C^k. Indeed, let φ_1 be given by the functions $x = x(u, v), y = y(u, v), z = z(u, v)$, and $\mathbf{h}$ by the functions $u = \varphi(\alpha, \beta), v = \psi(\alpha, \beta)$. Then φ_P is given by the functions

$$x = x(u(\alpha, \beta), v(\alpha, \beta)), \quad y = y(u(\alpha, \beta), v(\alpha, \beta)), \quad z = z(u(\alpha, \beta), v(\alpha, \beta)),$$

and in view of well-known theorems from analysis, φ_P belongs to class C^k.

To show the regularity of this parameterization, we have

$$\vec{r}_\alpha = \vec{r}_u \varphi_\alpha + \vec{r}_v \psi_\alpha, \quad \vec{r}_\beta = \vec{r}_u \varphi_\beta + \vec{r}_v \psi_\beta.$$

Thus

$$|\vec{r}_\alpha \times \vec{r}_\beta| = |\vec{r}_u \times \vec{r}_v| \cdot |\varphi_\alpha \psi_\beta - \varphi_\beta \psi_\alpha| = |\vec{r}_u \times \vec{r}_v| \cdot \det J \neq 0,$$

which completes the proof.

Sometimes, it appears necessary to introduce new coordinates on a surface so that the curves from a given family become the coordinate curves. Often, we arrive at the following situation. In a given coordinate neighborhood W with coordinates u, v, let two first-order differential equations are defined

$$A_1(u, v)\, du + B_1(u, v)\, dv = 0, \quad A_2(u, v)\, du + B_2(u, v)\, dv = 0, \quad (2.1)$$

and we wish to introduce a new coordinate system so that the integral curves of equation (2.1) will become the coordinate curves. The following lemma gives us a sufficient condition for the existence of such a parameterization.

Lemma 2.1.1. If at the point $P_0(u_0, v_0)$ the determinant

$$\det J = \begin{vmatrix} A_1(u_0, v_0) & B_1(u_0, v_0) \\ A_2(u_0, v_0) & B_2(u_0, v_0) \end{vmatrix}$$

is nonzero, then in some neighborhood of P_0 one can introduce coordinates such that the integral curves of equation (2.1) will be the coordinate curves.

Proof. Rewrite (2.1) in the form of two systems of ordinary differential equations

$$\begin{cases} \frac{du}{dt} = -B_1(u, v), \\ \frac{dv}{dt} = A_1(u, v), \end{cases} \quad (2.2)$$

$$\begin{cases} \frac{du}{dt} = -B_2(u, v), \\ \frac{du}{dt} = A_2(u, v). \end{cases} \quad (2.3)$$

Let γ_0 be the integral curve of system (2.2) passing through P_0, and let its equations be given by the functions $u = \varphi_1(t)$, $v = \psi_1(t)$. Then the functions $\varphi_1(t)$, $\psi_1(t)$ satisfy the equations

$$\begin{cases} \frac{d\varphi_1}{dt} = -B_1(\varphi_1, \psi_1), \\ \frac{d\psi_1}{dt} = A_1(\varphi_1, \psi_1), \end{cases} \quad \text{and} \quad \varphi_1(0) = u_0, \quad \psi_1(0) = v_0. \quad (2.4)$$

Analogously, let σ_0 be the integral curve of system (2.3) passing through the point P_0, and let $u = \varphi_2(\tau)$ and $v = \psi_2(\tau)$ be its equations. Then

$$\begin{cases} \frac{d\varphi_2}{d\tau} = -B_2(\varphi_2, \psi_2), \\ \frac{d\psi_2}{d\tau} = A_2(\varphi_2, \psi_2) \end{cases} \quad \text{and} \quad \varphi_2(0) = u_0, \quad \psi_2(0) = v_0. \quad (2.5)$$

Denote by $\gamma_\tau(\alpha)$ the integral curve of (2.2) passing through the point $\sigma_0(\tau)$, and by $\sigma_t(\beta)$ the integral curve of (2.3) passing through $\gamma_0(t)$. Let $P(t, \tau)$ be the point of intersection of $\gamma_\tau(\alpha)$ and $\sigma_t(\beta)$. Since the coordinates of $P(t, \tau)$ in W are u, v, then we obtain a map $u = u(t, \tau)$, $v = v(t, \tau)$. First, we prove that for sufficiently small t and τ the point $P(t, \tau)$ is uniquely defined. This means that we must prove the existence and uniqueness of a solution of the system of equations

$$\begin{cases} F_1(\alpha, \beta, t, \tau) = f_1(\alpha, \tau) - f_2(\beta, t) = 0, \\ F_2(\alpha, \beta, t, \tau) = h_1(\alpha, \tau) - h_2(\beta, t) = 0, \end{cases} \quad (2.6)$$

where $u = f_1(\alpha, \tau)$, $v = h_1(\alpha, \tau)$ are the equations of the curve $\gamma_\tau(\alpha)$, and $u = f_2(\beta, t)$, $v = h_2(\beta, t)$ are the equations of the curve $\sigma_t(\beta)$. From the definition of curves σ_0, γ_0, $\gamma_\tau(\alpha)$, and $\sigma_t(\beta)$ follow the equalities $f_1(0, 0) = u_0 \pm f_2(0, 0)$, $h_1(0, 0) = v_0 \pm h_2(0, 0)$. These equalities show us that the solution of the system (2.6) exists for $t = \tau = 0$. We now calculate the determinant $\det J_1 = \begin{vmatrix} F_{1,\alpha} & F_{1,\beta} \\ F_{2,\alpha} & F_{2,\beta} \end{vmatrix} \neq 0$ at the point $(0, 0)$:

$$F_{1,\alpha} = \frac{\partial f_1}{\partial \alpha}(0, 0) = \frac{\partial f_1}{\partial \alpha}(\alpha, 0)|_{\alpha=0} = \frac{\partial \varphi_1}{\partial \alpha}|_{\alpha=0} = -B_1(u_0, v_0),$$

$$F_{1,\beta} = \frac{\partial f_2}{\partial \beta}(0, 0) = -\frac{\partial f_2}{\partial \beta}(\beta, 0)|_{\beta=0} = \frac{\partial \varphi_2}{\partial \beta}|_{\beta=0} = -B_2(u_0, v_0),$$

$$F_{2,\alpha} = \frac{\partial h_1}{\partial \alpha}(0, 0) = \frac{\partial h_1}{\partial \alpha}(\alpha, 0)|_{\alpha=0} = \frac{\partial \psi_1}{\partial \alpha}|_{\alpha=0} = A_1(u_0, v_0),$$

$$F_{2,\beta} = \frac{\partial h_2}{\partial \beta}(0, 0) = -\frac{\partial h_2}{\partial \beta}(\beta, 0)|_{\beta=0} = \frac{\partial \psi_2}{\partial \beta}|_{\beta=0} = A_2(u_0, v_0). \quad (2.7)$$

From (2.7) we obtain

$$\det J_1 = -\det J \neq 0. \quad (2.8)$$

From (2.8) and the implicit function theorem, we deduce the existence of a real $\delta > 0$ such that for $t^2 + \tau^2 < \delta^2$ the functions $\alpha = \alpha(t, \tau)$ and $\beta = \beta(t, \tau)$ are defined and differentiable. Note also that

$$\alpha(t, 0) = t, \quad \beta(0, \tau) = \tau. \quad (2.9)$$

The functions $u = u(t, \tau)$ and $v = v(t, \tau)$ are determined by the formulas

$$u = u(t, \tau) = f_1(\alpha(t, \tau), \tau) = f_2(\beta(t, \tau), \tau),$$

$$v = v(t, \tau) = h_1(\alpha(t, \tau), \tau) = h_2(\beta(t, \tau), \tau). \quad (2.10)$$

The value of the determinant $\det J_2 = \begin{vmatrix} \frac{\partial u}{\partial t} & \frac{\partial u}{\partial \tau} \\ \frac{\partial v}{\partial t} & \frac{\partial v}{\partial \tau} \end{vmatrix}$ at the point $t = \tau = 0$ is equal to $- \det J$, and hence $\det J_2 \neq 0$. In fact, from (2.10), (2.9), and (2.7) it follows that

$$\frac{\partial u}{\partial t} = \frac{\partial f_1}{\partial \alpha}(0,0) \cdot \left.\frac{\partial \alpha}{\partial t}\right|_{t=0} = -B_1, \qquad \frac{\partial u}{\partial \tau} = \frac{\partial f_2}{\partial \beta}(0,0) \cdot \left.\frac{\partial \beta}{\partial \tau}\right|_{\tau=0} = -B_2,$$

$$\frac{\partial v}{\partial t} = \frac{\partial h_1}{\partial \alpha}(0,0) \cdot \left.\frac{\partial \alpha}{\partial t}\right|_{t=0} = A_1, \qquad \frac{\partial v}{\partial \tau} = \frac{\partial h_2}{\partial \beta}(0,0) \cdot \left.\frac{\partial \beta}{\partial \tau}\right|_{\tau=0} = A_2.$$

We apply again the inverse function theorem and obtain that the map $u = u(t, \tau)$, $v = v(t, \tau)$ induces new coordinates t and τ, for which $t = \text{const}$, $\tau = \tau$ and $t = t$, $\tau = \text{const}$ are integral curves of equations (2.2) and (2.3). $\qquad \square$

We prove one more lemma.

Lemma 2.1.2. Let $\vec{\lambda} = \lambda^1 \vec{r}_u + \lambda^2 \vec{r}_v$ and $\vec{\mu} = \mu^1 \vec{r}_u + \mu^2 \vec{r}_v$ be two nonparallel vectors at the point $P_0(u_0, v_0)$. Then there is a coordinate system ξ, η such that P is the point $(0, 0)$ and $\vec{r}_\xi = \vec{\lambda}$, $\vec{r}_\eta = \vec{\mu}$.

Proof. Let the functions $u = u(\xi, \eta)$ and $v = v(\xi, \eta)$ be defined by the formulas

$$u = \lambda^1 \xi + \mu^1 \eta + u_0, \qquad v = \lambda^2 \xi + \mu^2 \eta + v_0.$$

Then the coordinates of the point P are $(0, 0)$, and

$$\vec{r}_\xi = \vec{r}_u \frac{\partial u}{\partial \xi} + \vec{r}_v \frac{\partial v}{\partial \xi} = \lambda^1 \vec{r}_u + \lambda^2 \vec{r}_v = \vec{\lambda},$$

$$\vec{r}_\eta = \vec{r}_u \frac{\partial u}{\partial \eta} + \vec{r}_v \frac{\partial v}{\partial \eta} = \mu^1 \vec{r}_u + \mu^2 \vec{r}_v = \vec{\mu}. \qquad \square$$

2.2 The Tangent Plane

Let $\gamma(t): u = u(t)$, $v = v(t)$ $(a \leq t \leq b)$ be some curve through the point $P = \gamma(0) = (u(0), v(0))$. A tangent vector $\dot{\gamma} = \vec{\tau}$ to γ at P can be written in the form $\dot{\gamma} = \vec{r}'_t = \vec{r}_u u' + \vec{r}_v v'$. From this formula we see that the tangent vector to any curve on the surface Φ through the point P belongs to the plane of vectors $\vec{r}_u(P)$ and $\vec{r}_v(P)$. This observation leads us to the following definition.

Definition 2.2.1. The plane through a point P of a regular surface Φ of class C^k $(k \geq 1)$ that is parallel to the vectors $\vec{r}_u(P)$ and $\vec{r}_v(P)$ is called the *tangent plane* to the surface Φ at P. The *normal to the surface* Φ is an orthogonal vector to the tangent plane of the surface at the point P.

The normal to the surface Φ at the point P will be denoted by $\vec{n}(P)$ and assumed to be a unit vector: $|\vec{n}(P)| = 1$. In particular, the normal can be identified with the vector $\frac{\vec{r}_u \times \vec{r}_v}{|\vec{r}_u \times \vec{r}_v|}$.

Example 2.2.1. The *Möbius strip* is obtained when we twist a thin strip of paper through a 180° angle and then glue its ends. Consider its regular parameterization

$$\vec{r}(u, v) = \left[\left(R - v \sin \frac{v}{2} \right) \sin u, \left(R - v \sin \frac{v}{2} \right) \cos u, v \cos \frac{v}{2} \right],$$

$$0 < u < 2\pi, \quad |v| < 1,$$

where $R > 1$ and $n = 1$. The unit normal $\vec{n}$ will coincide with its opposite vector when one twists the path $u = t$, $v = 0$ from 0 to 2π. This surface is *nonorientable* for odd n, and it has only one side. For even n we twist a thin strip of paper through a $180°n$ angle, and the surface obtained is *orientable*, is homeomorphic to the cylinder, and has two sides. In other words, a regular surface Φ in $\mathbb{R}^3$ is *orientable* if and only if there is a differentiable field of unit normals $\vec{n} \colon \Phi \to \mathbb{R}^3$ on Φ. A *differentiable field of unit normals* on Φ is a differentiable map $\vec{n} \colon \Phi \to \mathbb{R}^3$ that associates with each $Q \in \Phi$ a unit normal vector $\vec{n}(Q)$ to Φ at Q.

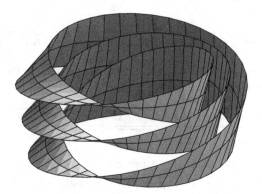

Figure 2.2. Möbius-type strips.

Let P be some point on the regular surface Φ, α the tangent plane to Φ at P, and Q an arbitrary point on Φ. Denote by d the distance between P and Q, and by h the distance from Q to the plane α. The following theorem contains the *geometrical characteristic of the tangent plane.*

Theorem 2.2.1. *If a regular surface Φ belongs to class C^1, then*

$$\lim_{Q \to P} \frac{h}{d} = \lim_{d \to 0} \frac{h}{d} = 0.$$

Proof. Let $\vec{r} = \vec{r}(u, v)$ be a parameterization of Φ, and let P have local coordinates (u_0, v_0). Then $d = |\vec{r}(u, v) - \vec{r}(u_0, v_0)|$, and $h = \langle \vec{r}(u, v) - \vec{r}(u_0, v_0), \vec{n} \rangle$. Set $\Delta u = u - u_0$, $\Delta v = v - v_0$. By Taylor's formula we have

$$\vec{r}(u, v) = \vec{r}(u_0, v_0) + \vec{r}_u(u_0, v_0)\Delta u + \vec{r}_v(u_0, v_0)\Delta v + \vec{o}(\sqrt{\Delta u^2 + \Delta v^2}).$$

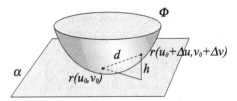

Figure 2.3. Geometrical characteristic of a tangent plane.

Thus,

$$d = |\vec{r}_u(u_0, v_0)\Delta u + \vec{r}_v(u_0, v_0)\Delta v + \vec{o}(\sqrt{\Delta u^2 + \Delta v^2})|,$$

$$h = \langle \vec{r}_u(u_0, v_0)\Delta u + \vec{r}_v(u_0, v_0)\Delta v + \vec{o}(\sqrt{\Delta u^2 + \Delta v^2}), \vec{n} \rangle$$

$$= |\langle \vec{o}(\sqrt{\Delta u^2 + \Delta v^2}), \vec{n} \rangle| = \vec{o}_1(\sqrt{\Delta u^2 + \Delta v^2}).$$

From the last formulas it follows that

$$\lim_{d \to 0} \frac{h}{d} = \lim_{d \to 0} \frac{\vec{o}_1(\sqrt{\Delta u^2 + \Delta v^2})}{|\vec{r}_u(u_0, v_0)\Delta u + \vec{r}_v(u_0, v_0)\Delta v + \vec{o}(\sqrt{\Delta u^2 + \Delta v^2})|}.$$

Divide the numerator and denominator of the above fraction by $\sqrt{\Delta u^2 + \Delta v^2}$. We shall prove that the expression

$$\frac{|\vec{r}_u(u_0, v_0)\Delta u + \vec{r}_v(u_0, v_0)\Delta v + \vec{o}(\sqrt{\Delta u^2 + \Delta v^2})|}{\sqrt{\Delta u^2 + \Delta v^2}}$$

is bounded from below by some positive number. For this, set $\xi = \frac{\Delta u}{\sqrt{\Delta u^2 + \Delta v^2}}$, $\eta = \frac{\Delta v}{\sqrt{\Delta u^2 + \Delta v^2}}$ and note that $\xi^2 + \eta^2 = 1$. We first estimate

$$|\vec{r}_u \xi + \vec{r}_v \eta|^2 = \langle \vec{r}_u, \vec{r}_u \rangle \xi^2 + 2\langle \vec{r}_u, \vec{r}_v \rangle \xi \eta + \langle \vec{r}_v, \vec{r}_v \rangle \eta^2.$$

We shall introduce the notation

$$E = \langle \vec{r}_u, \vec{r}_u \rangle, \quad F = \langle \vec{r}_u, \vec{r}_v \rangle, \quad G = \langle \vec{r}_v, \vec{r}_v \rangle$$

and calculate

$$EG - F^2 = |\vec{r}_u|^2 \cdot |\vec{r}_v|^2 - |\vec{r}_u|^2 \cdot |\vec{r}_v|^2 \cos^2 \varphi = |\vec{r}_u|^2 \cdot |\vec{r}_v|^2 \sin^2 \varphi = |\vec{r}_u \times \vec{r}_v|^2.$$

Here φ represents the angle between the vectors $\vec{r}_u$ and $\vec{r}_v$. Since the surface Φ is regular, $|\vec{r}_u \times \vec{r}_v| \neq 0$ and consequently, $EG - F^2 > 0$. Hence, the fundamental form $E\xi^2 + 2F\xi\eta + G\eta^2$ is positive definite. But then

$$\min_{\xi^2 + \eta^2 = 1} |E\xi^2 + 2F\xi\eta + G\eta^2| = a^2 > 0.$$

From the last inequality it follows that

$$\frac{|\vec{r}_u \Delta u + \vec{r}_v \Delta v + \vec{o}(\sqrt{\Delta u^2 + \Delta v^2})|}{\sqrt{\Delta u^2 + \Delta v^2}} \geq \frac{a}{2}$$

if $\sqrt{\Delta u^2 + \Delta v^2}$ is sufficiently small. Thus, we have that the denominator of the fraction is bounded from below by a positive number, and the numerator tends to zero. Hence, $\lim_{d \to 0} \frac{h}{d} = 0$, which completes the proof. □

Next, denote by $T\Phi_P$ the tangent plane to Φ at P. The vectors $\vec{r}_u$, $\vec{r}_v$ are linearly independent and hence form a vector space basis of $T\Phi_P$ called a *local basis of a surface* Φ *at a point* P.

Definition 2.2.2. Let α be some plane in the space $\mathbb{R}^3$, and $\vec{e}$ be a unit vector orthogonal to it. The pair $(\alpha, \vec{e})$ is called an *orientable plane*.

This definition is motivated by the fact that an orientation on a plane α is determined by a given orientation on $\mathbb{R}^3$ and a vector $\vec{e}$. And conversely, from an orientation on α and on $\mathbb{R}^3$ we can reconstruct a vector $\vec{e}$. Each orientable plane in the space generates in $\mathbb{R}^3$ a *height function* f_α *relative to a unit vector* $\vec{e}$, defined by the formula

$$f_\alpha(P) = \langle \overrightarrow{P_0 P}, \vec{e} \rangle,$$

where $P_0 \in \alpha$ is an arbitrary point. Obviously, f_α does not depend on the choice of P_0.

Exercise 2.2.1. Prove that if Q is a critical point of the height function f_α restricted to a regular surface Φ, then the tangent plane $T\Phi_Q$ is parallel to the plane α.

Exercise 2.2.2. Prove that for almost every plane α that intersects a regular surface Φ of class C^k, the intersection $\alpha \cap \Phi$ is a regular curve of class C^k.

2.2.1 Formulas for Calculations

The surface is given by parametric equations.

$$\vec{r} = \vec{r}(u, v) = x(u, v)\vec{i} + y(u, v)\vec{j} + z(u, v)\vec{k}.$$

Then

$$\vec{r}_u = x_u \vec{i} + y_u \vec{j} + z_u \vec{k}, \quad \vec{r}_v = x_v \vec{i} + y_v \vec{j} + z_v \vec{k}.$$

Thus the equation of the tangent plane to a surface at the point (u_0, v_0) is written in the form

$$(y_u z_v - z_u y_v) \cdot (X - x(u_0, v_0)) + (z_u x_v - x_u z_v) \cdot (Y - y(u_0, v_0))$$
$$+ (x_u y_v - y_u x_v) \cdot (Z - z(u_0, v_0)) = 0,$$

where X, Y, Z are coordinates of an arbitrary point in the plane, and all derivatives are calculated at the point (u_0, v_0).

Explicitly given surface. $x = x, y = y, z = f(x, y)$.

$$\vec{r}_x = \vec{i} + f_x\vec{k}, \quad \vec{r}_y = \vec{j} + f_y\vec{k}, \quad \vec{r}_x \times \vec{r}_y = -f_x\vec{i} - f_y\vec{j} + \vec{k}.$$

Let the coordinates of the point of tangency be $(x_0, y_0, z_0) = f(x_0, y_0)$. Then the equation of the tangent plane takes the form

$$Z - z_0 = f_x(x_0, y_0)(X - x_0) + f_y(x_0, y_0)(Y - y_0).$$

Implicitly given surface. $H(x, y, z) = 0$. Let (x_0, y_0, z_0) be the point on a surface. The equation of the tangent plane to the surface at this point has the form

$$H_x(x_0, y_0, z_0)(X - x_0) + H_y(x_0, y_0, z_0)(Y - y_0) + H_z(x_0, y_0, z_0)(Z - z_0) = 0.$$

2.3 First Fundamental Form of a Surface

Let P be an arbitrary point on a regular surface Φ. In the tangent plane $T\Phi_P$ we take two vectors $\vec{\lambda}$ and $\vec{\mu}$. Let λ^1, λ^2 and μ^1, μ^2 be coordinates of these vectors $\vec{\lambda}$ and $\vec{\mu}$ in a local basis $\vec{r}_u(P)$ and $\vec{r}_v(P)$. Find a formula for calculation of a scalar product of $\vec{\lambda}$ and $\vec{\mu}$ in terms of coordinates of these vectors in a local basis. Using our above notation $E = \langle \vec{r}_u, \vec{r}_u \rangle$, $F = \langle \vec{r}_u, \vec{r}_v \rangle$, $G = \langle \vec{r}_v, \vec{r}_v \rangle$, we obtain

$$\langle \vec{\lambda}, \vec{\mu} \rangle = E\lambda^1\mu^1 + F(\lambda^1\mu^2 + \lambda^2\mu^1) + G\lambda^2\mu^2.$$

In this way a scalar product generates on the surface Φ (in each tangent plane to Φ) a field of symmetric bilinear forms

$$I(\vec{\lambda}, \vec{\mu}) = E\lambda^1\mu^1 + F(\lambda^1\mu^2 + \lambda^2\mu^1) + G\lambda^2\mu^2.$$

In particular, the *first fundamental form of a surface* is defined as

$$I(\vec{\lambda}) = I(\vec{\lambda}, \vec{\lambda}) = E(\lambda^1)^2 + 2F\lambda^1\lambda^2 + G(\lambda^2)^2.$$

2.3.1 Length of a Curve on a Surface

Now let some smooth curve γ lie on a surface Φ, and let $u = u(t), v = v(t)$ $(a \leq t \leq b)$ be its equations in local coordinates. Find a formula for the length $l(\gamma)$ of γ. By the formula for the length of a curve, see Section 1.4, we have

$$l(\gamma) = \int_a^b |\vec{r}'|\, dt = \int_a^b \sqrt{I(\vec{r}')}\, dt = \int_a^b \sqrt{E(u')^2 + 2Fu'v' + G(v')^2}\, dt. \quad (2.11)$$

The formula (2.11) can be written in the form

$$l(\gamma) = \int_a^b \sqrt{E\, du^2 + 2F\, du\, dv + G\, dv^2}, \quad\quad\quad (2.12)$$

and the first fundamental form itself in the form

$$ds^2 = E\,du^2 + 2F\,du\,dv + G\,dv^2.$$

In this case, ds is said to be an *element of arc length*. This type of notation for the first fundamental form is *classical*. The formula (2.11) shows us that knowledge of the coefficients of the first fundamental form and the equations of a curve in local coordinates allow us to calculate the length of a curve.

Remark 2.3.1. If the equations of a surface are given, then in principle, we possess complete knowledge about the surface and its geometric properties. If the equation of a surface is unknown, but only the first fundamental form is given, then we of course do not possess all the information about the geometric properties of the surface. However, knowing the first fundamental form of a surface yields information about some geometric properties of the surface: we can define and study such geometric notions as *length of a curve, area of a region, shortest paths and geodesics, and Gaussian curvature of a surface*. Those geometric properties and objects that can be determined only in terms of the first fundamental form of a surface are called the *intrinsic geometric properties*, and the collection of these geometric properties and objects forms the subject of *intrinsic geometry of a surface*. In other words, one can say that the *intrinsic geometry of a surface studies such of its properties that do not depend on the shape of the surface, but depend only on measurements that we can carry out while staying on a surface itself*. The intrinsic geometry of a surface is the subject of Chapter 3. Now we shall define and study only the simplest notions related to the first fundamental form.

2.3.2 Metric on a Surface

The formula (2.11) was deduced for the case that a curve γ lies entirely in one coordinate neighborhood. But if a curve $\gamma(t)$ does not lie entirely in one coordinate neighborhood, then we divide it into a finite number of arcs, each of them lying in one coordinate neighborhood. Calculating the length of each of the arcs obtained by the formula (2.11) and summing their values, we obtain the length of the whole curve.

Lemma 2.3.1. If a regular surface Φ belongs to class C^1, then every two of its points can be joined by rectifiable (piecewise smooth) curve.

Proof. Let P and Q be two arbitrary points on the surface. In view of the connectedness of Φ, there is a continuous curve $\sigma(t)$ $(0 \le t \le 1)$ with endpoints $\sigma(0) = P$ and $\sigma(1) = Q$. Moreover, from the compactness of the set $\sigma(t)$ $(0 \le t \le 1)$ there follows the existence of a finite number of arcs σ_i $(t_i \le t \le t_{i+1})$, $i = 1, \ldots, n$, $t_1 = 0$, $t_n = 1$ of the curve σ, and coordinate neighborhoods W_i, $i = 1, \ldots, n$, such that $\sigma_i \subset W_i$. Let (u_i, v_i) and (u_{i+1}, v_{i+1}) be coordinates of the points $\sigma(t_i)$ and $\sigma(t_{i+1})$ in W_i. Take the curve γ_i defined by the equations

$$\begin{cases} u = u_i + \frac{t-t_i}{t_{i+1}-t_i}(u_{i+1} - u_i), \\ v = v_i + \frac{t-t_i}{t_{i+1}-t_i}(v_{i+1} - v_i), \end{cases} \qquad t_i \le t \le t_{i+1}, \quad i = 1, \dots, n,$$

and γ, composed of the arcs γ_i. Then γ is a piecewise smooth (rectifiable) curve and $\gamma(0) = P$, $\gamma(1) = Q$. □

Lemma 2.3.1 gives us the possibility to introduce the following definition.

Definition 2.3.1. The *distance* ρ on a surface Φ between the points P and Q is the greatest lower bound of the lengths of all curves on Φ with endpoints P and Q.

In other words, if $\mathbf{\Gamma}(P, Q)$ is the *set of all rectifiable curves on a surface Φ with endpoints P and Q*, then:

$$\rho(P, Q) = \inf_{\gamma \in \mathbf{\Gamma}(P, QP_1)} l(\gamma).$$

Lemma 2.3.1 says that $\mathbf{\Gamma}(P, Q)$ is nonempty, and that the nonnegative function $\rho(P, Q)$ is well-defined. The function $\rho(P, Q)$ has all usual properties of the distance function:

(1) $\rho(P, Q) = \rho(Q, P)$,

(2) $\rho(P, Q) + \rho(Q, R) \ge \rho(P, R)$,

(3) $\rho(P, Q) = 0$ if and only if $P = Q$.

Note also that the topology induced by this metric coincides with the topology induced from $\mathbb{R}^3$.

Introduction of such a metric on Φ allows us to give the following simple definitions.

Definition 2.3.2. A surface Φ is called *complete* if (Φ, ρ) is a complete metric space.

Definition 2.3.3. A surface Φ is called *closed (compact)* if (Φ, ρ) is a compact metric space.

2.3.3 Isometric Surfaces

Definition 2.3.4. Two regular surfaces Φ_1 and Φ_2 of class C^1 are called *isometric* if there is a map $\mathbf{h} \colon \Phi_1 \to \Phi_2$ preserving the length of every rectifiable curve. In other words, if γ_1 is a rectifiable curve on a surface Φ_1, and $\gamma_2 = \mathbf{h}(\gamma_1)$ is also rectifiable and $l(\gamma_1) = l(\gamma_2)$ for every $\gamma_1 \subset \Phi_1$, the map $\mathbf{h}$ is called an *isometry*, and the surfaces Φ_1 and Φ_2 are called *isometric* to each other.

Obviously, we can state this more simply: *Two surfaces Φ_1 and Φ_2 are isometric if they are isometric (to each other) as the metric spaces (Φ_1, ρ_1) and (Φ_2, ρ_2).*

For isometric surfaces we have the following theorem:

Theorem 2.3.1. *If regular surfaces Φ_1 and Φ_2 can be parameterized so that their first fundamental forms coincide, then these surfaces are isometric. The isometry* **h** *is defined by a one-to-one correspondence of the points with equal coordinates. Conversely, if Φ_1 and Φ_2 are isometric, then they can be parameterized so that their first fundamental forms coincide.*

Proof. The first part of the theorem's statement is obvious, and its second part will be proved in Chapter 3. □

Note also that isometric surfaces do not necessarily coincide under rigid motion. The simplest example is that a parabolic cylinder is isometric to a plane (verify).

2.3.4 Angle between the Curves on a Surface

Let $u = u_1(t)$, $v = v_1(t)$ and $u = u_2(t)$, $v = v_2(t)$ be equations of two regular curves γ_1 and γ_2 on a surface Φ of class C^1, and let γ_1 and γ_2 have a common point $\gamma_1(t_0) = \gamma_2(\tau_0)$. Then at this point it is possible to define the angle between the curves γ_1 and γ_2 as the angle between their tangent vectors $\vec{\tau}_1 = \gamma'_1$ and $\vec{\tau}_2 = \gamma'_2$. Since the coordinates of a vector $\vec{\tau}_1$ are $(u'_1(t_0), v'_1(t_0))$, and those of $\vec{\tau}_2$ are $(u'_1(\tau_0), v'_1(\tau_0))$, then

$$\cos \varphi = \frac{I(\vec{\tau}_1, \vec{\tau}_2)}{\sqrt{I(\vec{\tau}_1)}\sqrt{I(\vec{\tau}_2)}}$$

$$= \frac{Eu'_1 u'_2 + F(u'_1 v'_2 + v'_1 u'_2) + Gv'_1 v'_2}{\sqrt{E\left(u'_1\right)^2 + 2Fu'_1 v'_1 + G(v'_1)^2}\sqrt{E\left(u'_2\right)^2 + 2Fu'_2 v'_2 + G(v'_2)^2}}.$$

Note that the angle φ_{12} between coordinate curves is given by:

$$\cos \varphi_{12} = \frac{Fu'_1 v'_2}{\sqrt{E\left(u'_1\right)^2}\sqrt{G\left(v'_2\right)^2}} = \frac{F}{\sqrt{EG}}.$$

In particular, the equality $F = 0$ means that the coordinate curves are orthogonal.

2.3.5 Area of a Region on a Surface

Let D be some region on a surface Φ that lies entirely in some coordinate neighborhood. Define its area $S(D)$ by the following formula:

$$S(D) = \iint_D \sqrt{E(u, v)G(u, v) - F^2(u, v)}\, du\, dv. \qquad (2.13)$$

The expression $dS = \sqrt{E(u, v)G(u, v) - F^2(u, v)}\, du\, dv$ is called an *area element of the surface* Φ.

It is very difficult to give a visual geometric definition of the area of a region, and then, starting from it, to deduce (2.13), as was evidently done for the notion of arc length. Thus we take (2.13) as the definition of area and make some remarks explaining this formula. Take on the surface Φ the curvilinear "parallelogram" $\sigma: u_0 \leq u \leq u_0 + \Delta u,\ v_0 \leq v \leq v_0 + \Delta v$, and on a plane $T\Phi_{(u_0, v_0)}$ the parallelogram $\bar{\sigma}$ spanned by the vectors $\vec{r}_u(u_0, v_0)\Delta u$ and $\vec{r}_v(u_0, v_0)\Delta v$. The area $\bar{\sigma}$ of the parallelogram is

$$S(\bar{\sigma}) = |\vec{r}_u \times \vec{r}_v|\Delta u \Delta v = \sqrt{EG - F^2}\,\Delta u \Delta v.$$

From visual considerations, it seems to be true that the area of the "parallelogram" σ is approximately equal to the area $\bar{\sigma}$, and the error, which we do, has a higher order than $\Delta u \Delta v$. Thus, assuming

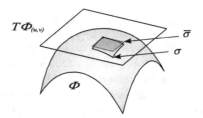

Figure 2.4. Area of a region on a surface.

$$S(D) = \lim \sum_i S(\sigma_i) = \lim \sum_i \left(\sqrt{EG - F^2}\,\Delta u_i \Delta v_i + \bar{o}_i(\Delta u_i \Delta v_i) \right),$$

where the limit is taken for *smaller and smaller subdivisions*, we obtain (2.13). If the whole region D does not lie in one coordinate neighborhood, then we divide it into sufficiently small parts so that each of them lies in one coordinate neighborhood, and define the area of D to be the sum of areas of its parts.

Example 2.3.1. A sphere of radius r, where φ and θ are *geographical coordinates* $u = \varphi, v = \theta$:

$$\begin{cases} x = r \cos \varphi \sin \theta, \\ y = r \sin \varphi \sin \theta, \\ z = r \cos \theta, \end{cases} \qquad 0 \leq \varphi < 2\pi, \quad 0 \leq \theta < \pi.$$

One can easily obtain

$$\vec{r}_u = -r \sin \varphi \sin \theta \vec{i} + r \cos \varphi \sin \theta \vec{j},$$
$$\vec{r}_v = r \cos \varphi \cos \theta \vec{i} + r \sin \varphi \cos \theta \vec{j} - r \sin \theta \vec{k},$$
$$E = r^2 \sin^2 \theta, \quad F = 0, \quad G = r^2 \ \rightarrow \ I(\vec{\lambda}) = r^2 \sin^2 \theta (\lambda^1)^2 + r^2(\lambda^2)^2.$$

Hence $\sqrt{EG - F^2} = r^2|\sin \theta|$, and the area of the sphere is

$$S = 2\pi r^2 \int_0^\pi |\sin \theta|\, d\theta = 4\pi r^2.$$

2.3.6 Formulas for Calculations

Surface given by the parametric equations. $x = x(u, v)$, $y = y(u, v)$, $z = z(u, v)$.

$$E = x_u^2 + y_u^2 + z_u^2, \quad F = x_u x_v + y_u y_v + z_u z_v, \quad G = x_v^2 + y_v^2 + z_v^2,$$

$$EG - F^2 = (x_u y_v - x_v y_u)^2 + (x_u z_v - x_v z_u)^2 + (y_u z_v - y_v z_u)^2,$$

$$S = \iint_D \sqrt{(x_u y_v - x_v y_u)^2 + (x_u z_v - x_v z_u)^2 + (y_u z_v - y_v z_u)^2} \, du \, dv.$$

Explicitly given surface. $z = f(x, y)$.

$$E = 1 + f_x^2, \quad F = f_x f_y, \quad G = 1 + f_y^2, \quad EG - F^2 = 1 + f_x^2 + f_y^2,$$

$$S = \iint_D \sqrt{1 + f_x^2 + f_y^2} \, dx \, dy.$$

2.4 Second Fundamental Form of a Surface

2.4.1 Normal curvature

Let P be some point on a regular surface Φ of class C^k ($k \geq 2$). Consider the plane $\Pi(P, \vec{\lambda})$ passing through a normal $\vec{n}$ to Φ at P and a vector $\vec{\lambda} \in T\Phi_P$. The intersection of this plane with Φ in some neighborhood of P is a regular curve γ of class C^k (see Exercise 2.2.2).

Denote by $\tilde{k}(P, \vec{\lambda})$ its curvature at P, and if $\tilde{k}(P, \vec{\lambda}) \neq 0$, then denote by $\vec{v}(P, \vec{\lambda})$ the principal normal vector of γ at P. Since γ lies in the plane $\Pi(P, \vec{\lambda})$, then $\vec{v}(P, \vec{\lambda}) = \pm \vec{n}(P)$. Define the real number $k(P, \vec{\lambda}) = \langle \vec{v}(P, \vec{\lambda}), \vec{n}(P) \rangle \tilde{k}(P, \vec{\lambda})$, i.e., assume that

$$k(P, \vec{\lambda}) = \begin{cases} \tilde{k}(P, \vec{\lambda}) & \text{if } \vec{v}(P, \vec{\lambda}) = \vec{n}(P), \\ 0 & \text{if } \tilde{k}(P, \vec{\lambda}) = 0, \\ -\tilde{k}(P, \vec{\lambda}) & \text{if } \vec{v}(P, \vec{\lambda}) = -\vec{n}(P). \end{cases}$$

Definition 2.4.1. The *normal curvature* of a surface Φ at a point P and in the direction $\vec{\lambda}$ is the real number $k(P, \vec{\lambda})$.

Obviously, the sign of $k(P, \vec{\lambda})$ depends on the choice of the direction of the normal $\vec{n}(P)$, and it changes generally with varying this direction. Thus the sign of $k(P, \vec{\lambda})$ itself has no geometric meaning. However, whether the sign varies or remains the same as the direction o f $\vec{\lambda}$ changes does have geometric significance. So, if $k(P, \vec{\lambda})$ has constant sign for all $\vec{\lambda} \in T\Phi_P$, then Φ in some neighborhood of P lies entirely on one side of its tangent plane $T\Phi_P$. But if $k(P, \vec{\lambda})$ changes sign, then Φ lies on both sides of $T\Phi_P$. In the first case, the point P is said to

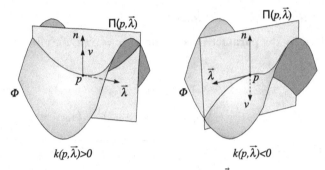

$$k(p,\vec{\lambda})>0 \qquad\qquad k(p,\vec{\lambda})<0$$

Figure 2.5. The sign of $k(P, \vec{\lambda})$.

be *elliptic* or a *point of convexity*, and in the second case, a *hyperbolic (saddle) point*. Thus the values of all normal curvatures at a point P on Φ allow us to form an opinion of the shape of the surface Φ in a sufficiently small neighborhood of P. Now let $\vec{r} = \vec{r}(u, v)$ be the equation of Φ, let $u = u(t)$, $v = v(t)$ be equations of a curve γ, and let t be a parameter, proportional to the arc length of the curve $\gamma : \vec{r}(t) = \vec{r}(u(t), v(t))$, with the properties

$$\gamma(0) = P, \quad \vec{\tau}(0) = \vec{\lambda} = \lambda^1 \vec{r}_u + \lambda^2 \vec{r}_v, \quad \lambda^1 = \frac{du}{dt}(0), \quad \lambda^2 = \frac{dv}{dt}(0).$$

We derive $k(P, \vec{\lambda})$:

$$k(P, \vec{\lambda}) = \tilde{k}(P, \vec{\lambda})\langle \vec{v}(P, \vec{\lambda}), \vec{n}(P)\rangle = \left|\frac{d^2\vec{r}}{dt^2}\right| \cdot \frac{1}{|\vec{\lambda}|^2}\langle \vec{v}(P, \vec{\lambda}), \vec{n}(P)\rangle, \qquad (2.14)$$

$$\frac{d^2\vec{r}}{dt^2} = \vec{r}_{uu}(\lambda^1)^2 + 2\vec{r}_{uv}\lambda^1\lambda^2 + \vec{r}_{vv}(\lambda^2)^2 + \vec{r}_u\frac{d^2u}{dt^2} + \vec{r}_v\frac{d^2v}{dt^2}. \qquad (2.15)$$

Substituting (2.15) in (2.14), we obtain

$$k(P, \vec{\lambda}) = \frac{1}{|\vec{\lambda}|^2}\left[\langle\vec{r}_{uu}, \vec{n}\rangle(\lambda^1)^2 + 2\langle\vec{r}_{uv}, \vec{n}\rangle\lambda^1\lambda^2 + \langle\vec{r}_{vv}, \vec{n}\rangle(\lambda^2)^2\right].$$

Introducing the notation

$$L = \langle\vec{r}_{uu}, \vec{n}\rangle, \quad M = \langle\vec{r}_{uv}, \vec{n}\rangle, \quad N = \langle\vec{r}_{vv}, \vec{n}\rangle,$$

we have

$$k(P, \vec{\lambda}) = \frac{L(\lambda^1)^2 + 2M\lambda^1\lambda^2 + N(\lambda^2)^2}{E(\lambda^1)^2 + 2F\lambda^1\lambda^2 + G(\lambda^2)^2} = \frac{II(\vec{\lambda})}{I(\vec{\lambda})}, \qquad (2.16)$$

where L, M, and N are derived at P. The quadratic form $II(\vec{\lambda})$ is called the *second fundamental form of the surface* Φ. Since P can be arbitrarily selected, $II(\vec{\lambda})$ is defined at each point on Φ. More exactly, the second fundamental form of Φ is defined on each of its tangent planes. Note that $II(\vec{\lambda})$ induces a field of symmetric bilinear forms

$$II(\vec{\lambda}, \vec{\mu}) = L\lambda^1\mu^1 + M(\lambda^1\mu^2 + \lambda^2\mu^1) + N\lambda^2\mu^2.$$

2.4.2 Formulas for Calculations

Surface given by the parametric equations. $x = x(u, v)$, $y = y(u, v)$, $z = z(u, v)$:

$$L = \langle \vec{r}_{uu}, \vec{n} \rangle = \frac{(\vec{r}_{uu} \cdot \vec{r}_u \cdot \vec{r}_v)}{|\vec{r}_u \times \vec{r}_v|} = \frac{1}{\sqrt{EG - F^2}} \begin{vmatrix} x_{uu} & y_{uu} & z_{uu} \\ x_u & y_u & z_u \\ x_v & y_v & z_v \end{vmatrix},$$

$$M = \langle \vec{r}_{uv}, \vec{n} \rangle = \frac{(\vec{r}_{uv} \cdot \vec{r}_u \cdot \vec{r}_v)}{|\vec{r}_u \times \vec{r}_v|} = \frac{1}{\sqrt{EG - F^2}} \begin{vmatrix} x_{uv} & y_{uv} & z_{uv} \\ x_u & y_u & z_u \\ x_v & y_v & z_v \end{vmatrix},$$

$$N = \langle \vec{r}_{vv}, \vec{n} \rangle = \frac{(\vec{r}_{vv} \cdot \vec{r}_u \cdot \vec{r}_v)}{|\vec{r}_u \times \vec{r}_v|} = \frac{1}{\sqrt{EG - F^2}} \begin{vmatrix} x_{vv} & y_{vv} & z_{vv} \\ x_u & y_u & z_u \\ x_v & y_v & z_v \end{vmatrix}.$$

Sometimes, for calculating L, M, and N it is more convenient to use the following formulas:

$$L = -\langle \vec{r}_u, \vec{n}_u \rangle, \quad M = -\langle \vec{r}_u, \vec{n}_v \rangle, \quad N = -\langle \vec{r}_v, \vec{n}_v \rangle, \tag{2.17}$$

which are obtained from the previous formulas by differentiation of the identities

$$\langle \vec{r}_u, \vec{n} \rangle = 0, \quad \langle \vec{r}_v, \vec{n} \rangle = 0.$$

From (2.16) it is seen that the second fundamental form admits the invariant definition

$$II(\vec{\lambda}) = -\left\langle \vec{\lambda}, \frac{d\vec{n}}{d\vec{\lambda}} \right\rangle.$$

In classical notation, the second fundamental form is defined by the formula

$$II(d\vec{r}) = -\langle d\vec{r}, d\vec{n} \rangle,$$

where the direction $d\vec{r}$ is defined by the ratio of differentials $du : dv$.

Explicitly given surface. $z = f(x, y)$.

$$L = \frac{f_{xx}}{\sqrt{1 + f_x^2 + f_y^2}}, \quad M = \frac{f_{xy}}{\sqrt{1 + f_x^2 + f_y^2}}, \quad N = \frac{f_{yy}}{\sqrt{1 + f_x^2 + f_y^2}}.$$

Example 2.4.1. We continue Example 2.3.1 with a sphere:

$$\vec{r}_{uu} = -r\cos\varphi\sin\theta\vec{i} - r\sin\varphi\sin\theta\vec{j},$$

$$\vec{r}_{uv} = -r\sin\varphi\cos\theta\vec{i} + r\cos\varphi\cos\theta\vec{j},$$

$$\vec{r}_{vv} = -r\cos\varphi\sin\theta\vec{i} - r\sin\varphi\sin\theta\vec{j} - r\cos\theta\vec{k},$$

$$L = \frac{-r\sin\varphi(r^2\sin^2\varphi\sin^2\theta + r^2\cos^2\varphi\sin^2\theta)}{2\sin\theta} = -r^2\sin^2\theta,$$

$$M = 0, \quad N = -r^2,$$

$$II(\vec{\lambda}) = -r^2\big[\sin^2\theta(\lambda^1)^2 + (\lambda^2)^2\big].$$

Example 2.4.2. Cylinder:
$$\begin{cases} x = r\cos v \\ y = r\sin v \\ z = \quad u \end{cases}, \quad II(\vec{\lambda}) = \tfrac{1}{r}(\lambda^1)^2.$$

Example 2.4.3. Helicoid:
$$\begin{cases} x = u\cos v \\ y = u\sin v \\ z = \quad cv \end{cases}, \quad E = 1, \quad F = 0, \quad G = u^2 + c^2,$$

$$L = N = 0, \quad M = \frac{c}{\sqrt{u^2 + c^2}}, \quad II(\vec{\lambda}) = \frac{c}{\sqrt{u^2 + c^2}}\lambda^1\lambda^2.$$

2.4.3 Meusnier's Theorem

Let γ_1 be an arbitrary C^2-regular curve on a regular surface Φ of class C^2 passing through the point P in the direction of the vector $\vec{\lambda}$. If a curvature k_1 of γ_1 at P differs from zero, then denote by $\vec{v}_1$ the principal normal vector of γ_1 at P. Define the *sign of curvature* k_1 analogously to previous considerations: the sign of k_1 is assumed to be equal to the sign of $\langle \vec{v}_1, \vec{n}(P) \rangle$ if $\langle \vec{v}_1, \vec{n}(P) \rangle \neq 0$, and denote by θ the angle between the vectors $\vec{n}$ and $\vec{v}_1$.

Theorem 2.4.1 (Meusnier). *The curvatures k_1 and $k(P, \vec{\lambda})$ are related by the formula*

$$k(P, \vec{\lambda}) = k_1 \cdot \cos\theta.$$

Proof. Let $u = u_1(t)$, $v = v_1(t)$ be equations of the curve γ_1; let t be a parameter proportional to the arc length of $\gamma_1 \colon \vec{r}(t) = \vec{r}(u_1(t), v_1(t))$ such that

$$\dot{\gamma}_1(0) = P, \quad \frac{du_1}{dt}(0) = \lambda^1, \quad \frac{dv_1}{dt}(0) = \lambda^2, \quad \vec{\lambda} = \lambda^1\vec{r}_u + \lambda^2\vec{r}_v.$$

Since

$$\frac{d^2\vec{r}}{dt^2} = \vec{r}_{uu}(\lambda^1)^2 + 2\vec{r}_{uv}\lambda^1\lambda^2 + \vec{r}_{vv}(\lambda^2)^2 + \vec{r}_u\frac{d^2u}{dt^2} + \vec{r}_v\frac{d^2v}{dt^2},$$

then

$$\left\langle \frac{d^2\vec{r}}{dt^2}, \vec{n} \right\rangle = II(\vec{\lambda}) = k(P, \vec{\lambda}) \cdot I(\vec{\lambda}). \tag{2.18}$$

On the other hand (see Section 1.6),

$$\frac{d^2\vec{r}}{dt^2} = |k_1| \cdot |\vec{\lambda}|^2 \cdot \vec{v}_1. \tag{2.19}$$

From (2.18), (2.19), and the definition of the sign of k_1, it follows that

$$k(P, \vec{\lambda}) = |k_1| \cdot \langle \vec{v}_1, \vec{n}(P) \rangle = k_1 \cdot \cos\theta \tag{2.20}$$

and the theorem is proved. $\qquad\qquad\qquad\qquad\qquad\qquad\qquad\qquad\square$

The formula (2.20) can be used to determine the normal curvature, as usual. In fact, $k(P, \vec{\lambda})$ does not depend on the choice of the curve γ, but only on choices of the vector $\vec{\lambda}$ and the direction of the normal $\vec{n}(P)$.

If $k(P, \vec{\lambda}) \neq 0$, then Meusnier's theorem has a beautiful geometric interpretation. Let $\Pi(P, \vec{\lambda}, \theta)$ be the plane through P that is parallel to $\vec{\lambda}$ and forms an angle θ ($0 \leq \theta < \pi/2$) with the plane $\Pi(P, \vec{\lambda})$. Denote by $\gamma(\vec{\lambda}, \theta)$ the curve of the intersection of $\Pi(P, \vec{\lambda}, \theta)$ and Φ, and by $k(\vec{\lambda}, \theta)$ its curvature. From P in the direction of the principal normal vector of the curve $\gamma(\vec{\lambda}, \theta)$, let us mark off a line segment equal to the curvature radius $R(\vec{\lambda}, \theta) = 1/k(\vec{\lambda}, \theta)$. Denote by $C(\vec{\lambda})$ the geometric locus of the points obtained. From Meusnier's theorem it follows that $C(\vec{\lambda})$ is a circle with diameter $d = 1/k_1(P, \vec{n})$; see Figure 2.6. Note that $C(\vec{\lambda})$ lies

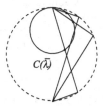

Figure 2.6. Meusnier's theorem: geometrical interpretation.

in the plane orthogonal to $\Pi(P, \vec{\lambda}, \theta)$, and the ends of its diameter are located at the points $\vec{r}(P)$ and $\vec{r}(P) + R(\vec{\lambda}, \theta)\vec{v}(P)$.

2.4.4 Principal Curvatures and Principal Vectors

Let P be an arbitrary point on a regular surface Φ of class C^2. We shall study the behavior of the normal curvature as a function of direction $\vec{\lambda}$:

$$k(P, \vec{\lambda}) = \frac{L(\lambda^1)^2 + 2M\lambda^1\lambda^2 + N(\lambda^2)^2}{E(\lambda^1)^2 + 2F\lambda^1\lambda^2 + G(\lambda^2)^2}.$$

Since all directions in a tangent plane $T\Phi_P$ form a compact set homeomorphic to a circle, $k(P, \vec{\lambda})$ has at least one minimum and one maximum, i.e., at least two extremal values.

Definition 2.4.2. A *principal curvature of a surface* Φ *at a point* P is an extremal value of a normal curvature function $k(P, \vec{\lambda})$ at a given point P on Φ. The directions (vectors) in $T\Phi_P$ for which $k(P, \vec{\lambda})$ takes its extremal values are *principal vectors of the surface* Φ *at the point* P.

We now derive the equations for deriving the principal curvatures and the principal vectors. Let a_0 be a principal curvature, $\vec{\lambda}_0$ the corresponding principal vector, $\vec{\lambda}_0 = \lambda_0^1 \vec{r}_u + \lambda_0^2 \vec{r}_v$, and $\vec{\lambda} = \lambda^1 \vec{r}_u + \lambda^2 \vec{r}_v$ an arbitrary vector in the plane $T\Phi_P$. Then the function

$$f(\lambda^1, \lambda^2, a_0) = II(\vec{\lambda}) - a_0 \cdot I(\vec{\lambda})$$

and its derivatives $\partial f / \partial \lambda^1$ and $\partial f / \partial \lambda^2$ take zero values for $\lambda^1 = \lambda_0^1$, $\lambda^2 = \lambda_0^2$. In this way, we obtain the following system of equations:

$$\begin{cases} II(\vec{\lambda}_0) - a_0 \cdot I(\vec{\lambda}_0) = 0, \\ II_{\lambda^1}(\vec{\lambda}_0) - a_0 \cdot I_{\lambda^1}(\vec{\lambda}_0) = 0, \\ II_{\lambda^2}(\vec{\lambda}_0) - a_0 \cdot I_{\lambda^2}(\vec{\lambda}_0) = 0, \end{cases} \tag{2.21}$$

or, in more detailed form,

$$\begin{cases} (L - a_0 E)\lambda_0^1 + (M - a_0 F)\lambda_0^2 = 0 \\ (M - a_0 F)\lambda_0^1 + (N - a_0 G)\lambda_0^2 = 0. \end{cases} \tag{2.22}$$

Since $\vec{\lambda}_0 \neq 0$, the system (2.22) has nonzero solution λ_0^1, λ_0^2, and hence

$$\begin{vmatrix} L - a_0 E & M - a_0 F \\ M - a_0 F & N - a_0 G \end{vmatrix} = 0,$$

or

$$(EG - F^2)a_0^2 - (EN + GL - 2MF)a_0 + LN - M^2 = 0. \tag{2.23}$$

Thus, we see that there exist not more than two principal curvatures. But if (2.23) has a unique solution, then $\min_{\vec{\lambda}} k(P, \vec{\lambda}) = \max_{\vec{\lambda}} k(P, \vec{\lambda})$, and $k(P, \vec{\lambda})$ does not depend on $\vec{\lambda}$. In this case any $\vec{\lambda} \in T\Phi_P$ is a principal vector. Denote by $k_1(P)$ and $k_2(P)$ the *principal curvatures of the surface* Φ *at the point* P; $k_1(P) \leq k_2(P)$. We now deduce the equation for deriving the principal vectors. The system (2.21) has nonzero solution $(1, -a_0)$. Thus

$$II_{\lambda^1}(\vec{\lambda}_0) \cdot I_{\lambda^2}(\vec{\lambda}_0) - II_{\lambda^2}(\vec{\lambda}_0) \cdot I_{\lambda^1}(\vec{\lambda}_0) = 0,$$

or in more detailed form,

$$(E\lambda_0^1 + F\lambda_0^2)(M\lambda_0^1 + N\lambda_0^2) - (F\lambda_0^1 + G\lambda_0^2)(L\lambda_0^1 + M\lambda_0^2) = 0.$$

The last equation can be written in a form more convenient for applications and memorization:

$$\begin{vmatrix} -(\lambda_0^2)^2 & \lambda_0^1\lambda_0^2 & -(\lambda_0^1)^2 \\ E & F & G \\ L & M & N \end{vmatrix} = 0. \qquad (2.24)$$

We return to (2.23), which defines the principal curvatures k_1 and k_2:

$$(EG - F^2)k^2 - (EN + GL - 2MF)k + LN - M^2 = 0. \qquad (2.25)$$

In the theory of two-dimensional surfaces in three-dimensional Euclidean space $\mathbb{R}^3$, the following invariants of a point P on a surface Φ are of great importance.

Definition 2.4.3. *The Gaussian* (or *complete*) *curvature of a surface,* $K(P)$, is defined by the formula

$$K(P) = k_1(P) \cdot k_2(P),$$

and the *mean curvature of a surface,* $H(P)$, is defined by the formula

$$H(P) = \frac{1}{2}(k_1(P) + k_2(P)).$$

Applying Viète's theorem to (2.25), we obtain

$$K(P) = \frac{LN - M^2}{EG - F^2}, \quad H(P) = \frac{EN + GL - 2FM}{2(EG - F^2)}. \qquad (2.26)$$

From (2.26) we see that if the Gaussian curvature of a surface Φ is positive at some point P, then the second fundamental form at this point is positive or negative definite, and then the normal curvature $k(P, \vec{\lambda})$ of Φ does not change its sign when $\vec{\lambda}$ varies. In this case, as we know, a surface Φ in some neighborhood of P lies entirely on one side of its tangent plane, and such a point is said to be *elliptic* or a *point of convexity*. Examples are points on an ellipsoid or a sphere of radius R: in the latter case, $K = 1/R^2$.

If $K(P) < 0$, then $k_1(P) < 0$ and $k_2(P) > 0$, and P is called a *hyperbolic (saddle) point*, and the surface lies on both sides of its tangent plane. The origin of the terms elliptic or hyperbolic point is the fact that in a neighborhood of an elliptic point the shape of the surface Φ is similar, to a high order of accuracy, to an elliptic paraboloid, and in a neighborhood of a hyperbolic point to a hyperbolic paraboloid.

A point where $K(P) = 0$ is said to be *parabolic (or cylindrical),* since in this case one of the principal curvatures is zero, which occurs at any point of a cylinder.

Figure 2.7. Elliptic, hyperbolic, and parabolic points on a surface.

The mean curvature of a surface plays a less important role in the theory of surfaces than Gaussian curvature. We make only one remark.

A surface Φ is *minimal* if the mean curvature vanishes at all points of Φ.

A minimal surface is characterized locally by the property of having minimal area among all surfaces with the same boundary contour.

2.4.5 Umbilics

Definition 2.4.4. A point P on a surface Φ is called an *umbilical point* or *umbilic* if the principal curvatures $k_1(P)$ and $k_2(P)$ are equal. If in addition the equality $k_1(P) = k_2(P) = 0$ holds, then P is called a *planar point*.

The notion of these points is explained by the fact that on a sphere of radius R the equality $k_1(P) = k_2(P) = \pm 1/R$ holds at every point P, and at every point on a plane, $k_1(P) = k_2(P) = 0$ holds. Later, we shall prove a theorem that *if all points on a surface Φ are umbilics, then Φ is a (open) domain of a sphere or a plane.*

The normal curvature at an umbilic does not depend on the direction, and thus the second fundamental form is proportional to the first fundamental form, and the coefficient of proportionality is the principal curvature of the surface Φ at this point: $L = kE, M = kF, N = kG$, where $k = k_1 = k_2$. The opposite is also true: if $\frac{L}{E} = \frac{M}{F} = \frac{N}{G}$, then P is an umbilic.

By the way, we have obtained the following theorem.

Theorem 2.4.2. *A point P on a regular surface Φ of class C^k ($k \geq 2$) is umbilic if and only if the equalities $\frac{L}{E} = \frac{M}{F} = \frac{N}{G}$ hold, and P is a planar point if and only if the equalities $L = M = N = 0$ hold.*

It can be shown that a connected surface consisting only of umbilical points is part of a sphere. The general ellipsoid has four umbilics; see also Problem 2.7.2. The Carathéodory conjecture that *a surface homeomorphic to a sphere has at least two umbilics* is still open.[5]

2.4.6 Orthogonality of Principal Vectors

If a point P on a surface Φ is not umbilic, then there exist exactly two principal vectors in the tangent plane $T\Phi_P$. It turns out that these directions are mutually orthogonal, and adjoint with respect to the second fundamental form.

Theorem 2.4.3. *If a point P on a regular surface Φ is not umbilic, and two vectors $\vec{\lambda}_1, \vec{\lambda}_2 \in T\Phi_P$ are parallel to the principal vectors of Φ at P, then $I(\vec{\lambda}_1, \vec{\lambda}_2) = II(\vec{\lambda}_1, \vec{\lambda}_2) = 0$.*

---

[5] A recent survey about umbilics on surfaces is Gutierrez, C. and Sotomayor, J., *Lines of curvature, umbilic points and Carathéodory conjecture.* Resen. Inst. Mat. Estat. Univ. Sao Paulo, Vol. 3, No. 3, 291–322, 1998. The proof for the analytical case can be found in Ivanov, V.V., *The analytic conjecture of Carathéodory*, Siberian Math. J., v. 43, 251–322, 2002.

Proof. We write down (2.21), which defines the principal vectors:

$$\begin{cases} II_{\lambda_1^1}(\vec{\lambda}_1) - k_1 \cdot I_{\lambda_1^1}(\vec{\lambda}_1) = 0, \\ II_{\lambda_1^2}(\vec{\lambda}_1) - k_1 \cdot I_{\lambda_1^2}(\vec{\lambda}_1) = 0, \end{cases} \tag{2.27}$$

$$\begin{cases} II_{\lambda_2^1}(\vec{\lambda}_2) - k_2 \cdot I_{\lambda_2^1}(\vec{\lambda}_2) = 0, \\ II_{\lambda_2^2}(\vec{\lambda}_2) - k_2 \cdot I_{\lambda_2^2}(\vec{\lambda}_2) = 0. \end{cases} \tag{2.28}$$

Multiplying the first equation of system (2.27) by λ_2^1 and the second equation by λ_2^2 and summing them, we obtain

$$2 \cdot II(\vec{\lambda}_1, \vec{\lambda}_2) - 2k_1 \cdot I(\vec{\lambda}_1, \vec{\lambda}_2) = 0. \tag{2.29}$$

Analogously, multiplying the first equation of system (2.28) by λ_1^1 and the second equation by λ_1^2 and summing them, we obtain

$$2 \cdot II(\vec{\lambda}_1, \vec{\lambda}_2) - 2k_2 \cdot I(\vec{\lambda}_1, \vec{\lambda}_2) = 0. \tag{2.30}$$

Now subtracting (2.30) from (2.29), and by taking into account that $k_2 - k_1 \neq 0$, we obtain

$$I(\vec{\lambda}_1, \vec{\lambda}_2) = \langle \vec{\lambda}_1, \vec{\lambda}_2 \rangle = 0. \tag{2.31}$$

From (2.29) and (2.31) follows $II(\vec{\lambda}_1, \vec{\lambda}_2) = 0$.   □

2.4.7 Euler's Formula

A formula first deduced by Euler allows us to derive the normal curvature of a surface Φ at a given point P and in an arbitrary direction from the known principal curvatures k_1 and k_2 of the surface at this point. Denote by $\vec{\lambda}_1 = \lambda_1^1 \vec{r}_u + \lambda_1^2 \vec{r}_v$ and $\vec{\lambda}_2 = \lambda_2^1 \vec{r}_u + \lambda_2^2 \vec{r}_v$ two mutually orthogonal unit vectors going along the principal directions. The existence of such vectors follows from Theorem 2.4.3 and the remark before Theorem 2.4.2. Denote by $\vec{\lambda}(\varphi)$ the vector that forms an angle φ $(0 \leq \varphi \leq 2\pi)$ with $\vec{\lambda}_1$.

Theorem 2.4.4. *For an arbitrary point P on a regular surface Φ of class C^k $(k \geq 2)$, the following equality (called Euler's formula) holds:*

$$k(P, \varphi) = k(P, \vec{\lambda}(\varphi)) = k_1 \cos^2 \varphi + k_2 \sin^2 \varphi.$$

Proof. Introduce a coordinate system (u, v) in a neighborhood of P such that $P(0, 0)$ and $\vec{r}_u(0, 0) = \vec{\lambda}_1$, $\vec{r}_v(0, 0) = \vec{\lambda}_2$. The existence of such coordinates follows from Lemma 2.1.2. In this coordinate system we have

$$\lambda_1^1 = 1, \quad \lambda_1^2 = 0, \quad \lambda_2^1 = 0, \quad \lambda_2^2 = 1, \quad \lambda^1(\varphi) = \cos\varphi, \quad \lambda^2(\varphi) = \sin\varphi,$$
$$E(0, 0) = G(0, 0) = 1, \quad F(0, 0) = M(0, 0) = 0.$$

All these equalities follow from the definition of E, F, G, but the last equality follows from Theorem 2.4.3, since $II(\vec{\lambda}_1, \vec{\lambda}_2) = M = 0$. Furthermore,

$$k(P, \varphi) = \frac{L \cos^2 \varphi + N \sin^2 \varphi}{\cos^2 \varphi + \sin^2 \varphi} = L \cos^2 \varphi + N \sin^2 \varphi.$$

On the other hand,

$$k_1 = k(P, 0) = L, \quad k_2 = k\left(P, \frac{\pi}{2}\right) = N.$$

Thus we obtain $k(P, \varphi) = k_1 \cos^2 \varphi + k_2 \sin^2 \varphi$. □

2.4.8 Rodrigues's Theorem

Theorem 2.4.5 (Rodrigues). *The derivative of a normal $\vec{n}(P)$ to a regular surface Φ of class C^k ($k \geq 2$) along some direction is parallel to it if and only if this direction is the principal vector of the surface at P and the coefficient of proportionality is equal to $-k$, where k is the principal curvature of Φ at P corresponding to this principal vector.*

Proof. Let $\vec{\lambda} = \lambda^1 \vec{r}_u + \lambda^2 \vec{r}_v$ be the vector that defines the direction under discussion. The derivative of the vector field $\vec{n}(P)$ along the direction $\vec{\lambda}$, in view of the definition of directional derivative, is written in the following form:

$$\frac{d\vec{n}}{d\vec{\lambda}} = \lambda^1 \vec{n}_u + \lambda^2 \vec{n}_v. \tag{2.32}$$

Let $\vec{\lambda}$ be a principal vector. Then by Theorem 2.4.3, there is a vector $\vec{\mu} = \mu^1 \vec{r}_u + \mu^2 \vec{r}_v$ such that the following equalities are satisfied:

$$I(\vec{\lambda}, \vec{\mu}) = II(\vec{\lambda}, \vec{\mu}) = 0.$$

The scalar product of (2.32) by the vector $\vec{\mu}$ gives us

$$\left\langle \frac{d\vec{n}}{d\vec{\lambda}}, \vec{\mu} \right\rangle = -II(\vec{\lambda}, \vec{\mu}) = 0. \tag{2.33}$$

Consequently, $\frac{d\vec{n}}{d\vec{\lambda}}$ is orthogonal to the vector $\vec{\mu}$. Hence it is collinear with $\vec{\lambda}$, and we have $\frac{d\vec{n}}{d\vec{\lambda}} = a\vec{\lambda}$. We now find the value of a. The scalar product of the last equation by the vector $\vec{\lambda}$ gives us $-II(\vec{\lambda}) = a \cdot I(\vec{\lambda})$, or $a = -\frac{II(\vec{\lambda})}{I(\vec{\lambda})} = -k$. Now let the equality

$$\frac{d\vec{n}}{d\vec{\lambda}} = a\vec{\lambda} \tag{2.34}$$

be given. As before, we obtain $a = -k$, and it remains to prove only that $\vec{\lambda}$ is parallel to the principal vector. From (2.32) and (2.34) follows the equality

$$\lambda^1 \vec{n}_u + \lambda^2 \vec{n}_v = (\lambda^1 \vec{r}_u + \lambda^2 \vec{r}_v)(-k).$$

Its scalar product, first by $\vec{r}_u$, and then by $\vec{r}_v$, gives us a system of equations

$$\begin{cases} -L\lambda^1 - M\lambda^2 = -kE\lambda^1 - kF\lambda^2, \\ -M\lambda^1 - N\lambda^2 = -kF\lambda^1 - kG\lambda^2, \end{cases}$$

which coincides with the system (2.22), and hence $\vec{\lambda}$ is a principal vector. The theorem is proved. □

Consider one application of Rodrigues's theorem. Solve the following problem.

Problem 2.4.1. If the normal curvature $k(P, \vec{\lambda})$ of a regular surface Φ of class C^k ($k \geq 3$) depends on neither P nor $\vec{\lambda}$, then Φ is an open connected domain on either a sphere or a plane.

Solution. Take an arbitrary point $P \in \Phi$. Select the direction of a normal $\vec{n}$ at P and on some neighborhood in such a way that the normal curvatures of the surface are positive; i.e., assume that $k(P, \vec{\lambda}) \equiv k_0 > 0$. Take an arbitrary direction $\vec{\lambda} \in T\Phi_P$ and denote by $\Pi(P, \vec{\lambda})$ the plane through a point P that is parallel to the vectors $\vec{n}$ and $\vec{\lambda}$, and by $\gamma(P, \vec{\lambda})$ the curve that appears as the intersection of Φ and a plane $\Pi(P, \vec{\lambda})$. Introduce on $\gamma(P, \vec{\lambda})$ the arc length parameterization with t counting from the point P. Denote by $\vec{\tau}(t)$, $\vec{v}(t)$, and $\vec{\beta}(t)$ a tangent vector, the principal normal vector and the binormal of $\gamma(P, \vec{\lambda})$. The curve $\gamma(P, \vec{\lambda})$ is a plane curve, and its torsion is zero. Thus from the Frenet formulas (see Section 1.9) we obtain

$$\vec{\tau}'(t) = k\vec{v}(t), \quad \vec{v}'(t) = -k\vec{\tau}(t), \quad \vec{\beta}'(t) = 0, \tag{2.35}$$

where $k(t)$ is the curvature of $\gamma(P, \vec{\lambda})$. On the other hand, from Rodrigues's theorem we have

$$\vec{n}'(t) = -k_0 \vec{\tau}(t). \tag{2.36}$$

Set $\vec{a}(t) = \vec{n}(t) - \vec{v}(t)$. Then $\vec{a}(0) = 0$, and since $\vec{a}(t)$ is orthogonal to $\vec{\tau}(t)$, then

$$\vec{a}(t) = c_1(t)\vec{\beta}(t) + c_2(t)\vec{v}(t), \tag{2.37}$$

where $c_1(t)$ and $c_2(t)$ are some differentiable functions. Thus, on the one hand, we have

$$\frac{d\vec{a}}{dt} = \frac{d\vec{n}}{dt} - \frac{d\vec{v}}{dt} = (k - k_0)\vec{\tau}, \tag{2.38}$$

but on the other hand,

$$\frac{d\vec{a}}{dt} = c_1'(t)\vec{\beta} + c_2'(t)\vec{v} + c_2(t)(-k\vec{\tau}). \tag{2.39}$$

From (2.38) and (2.39) we obtain

$$(k - k_0)\vec{\tau} = c_1'(t)\vec{\beta} + c_2'(t)\vec{v} - c_2(t)k\vec{\tau}, \tag{2.40}$$

or

$$k + c_2 k - k_0 = 0, \quad c_1'(t) = 0, \quad c_2'(t) = 0,$$

and using (2.37): $c_1(t) = c_2(t) = 0$. Hence $k(t) = k_0$. Consequently, $\gamma(P, \vec{\lambda})$ is a circular arc of radius $\frac{1}{k_0}$. Since P and $\vec{\lambda}$ are arbitrarily chosen, the problem is solved. $\square$

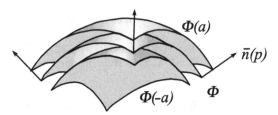

Figure 2.8. Parallel surfaces.

The study of parallel surfaces in the next section is based in an essential way on Rodrigues's theorem.

2.4.9 Parallel Surfaces

Let Φ be a regular surface of class C^k ($k \geq 3$), and $\vec{n}(P)$ a normal vector field on Φ. For arbitrary real a we shall construct a surface $\Phi(a)$ by marking off a line segment of length $|a|$ from each point $P \in \Phi$ in the direction of the normal $\vec{n}(P)$ if $a > 0$, and in the direction $-\vec{n}(P)$ if $a < 0$. The surface $\Phi(a)$ is said to be a *parallel surface* to the surface Φ. Obviously, the properties of a parallel surface $\Phi(a)$ are determined by the properties of Φ and by the value of the real number a.

Let us formulate the main theorem. Denote by $\varphi(P)$ the map from Φ to $\Phi(a)$ induced by the construction of $\Phi(a)$, and by $k_1(P, a)$, $k_2(P, a)$, and $\vec{n}(P, a)$ the principal curvatures and the normal to $\Phi(a)$ at the point $\varphi(P) \in \Phi(a)$, and set

$$R_1(P, a) = \frac{1}{k_1(P, a)}, \quad R_2(P, a) = \frac{1}{k_2(P, a)}.$$

Theorem 2.4.6. *If $\frac{1}{a} \neq k_1(P) = k(P, 0)$, $\frac{1}{a} \neq k_2(P) = k(P, 0)$, then a surface $\Phi(a)$ is regular at the point $\varphi(P)$, the normal $\vec{n}(P, a)$ coincides with $\vec{n}(P)$, the principal vectors of Φ at P are transformed to the principal vectors of $\Phi(a)$ at the point $\varphi(P)$, and the principal curvatures $k_1(P, a)$ and $k_2(P, a)$ are expressed by the formulas*

$$k_1(P, a) = \frac{k_1(P)}{1 - ak_1(P)}, \quad k_2(P, a) = \frac{k_2(P)}{1 - ak_2(P)} \tag{2.41}$$

or

$$R_1(P, a) = R_1(P) - a, \qquad R_2(P, a) = R_2(P) - a. \qquad (2.42)$$

Proof. Introduce a parameterization $\vec{r} = \vec{\rho}(u, v)$ of the surface Φ in a neighborhood of P such that P has nonzero coordinates and the vectors $\vec{\rho}_u(0, 0)$ and $\vec{\rho}_v(0, 0)$ are parallel to the principal vectors of Φ at P. The equation of the surface $\Phi(a)$ can be written as

$$\vec{r} = \vec{r}(u, v) = \vec{\rho}(u, v) + a\vec{n}(u, v).$$

We derive $\vec{r}_u \times \vec{r}_v$ at P

$$\vec{r}_u = \vec{\rho}_u + a\vec{n}_u, \qquad \vec{r}_v = \vec{\rho}_v + a\vec{n}_v.$$

From Rodrigues's theorem it follows that

$$\vec{r}_u = (1 - ak_1)\vec{\rho}_u, \qquad \vec{r}_v = (1 - ak_2)\vec{\rho}_v. \qquad (2.43)$$

From (2.43) and the conditions of the theorem we have

$$\vec{r}_u \times \vec{r}_v = (1 - ak_1)(1 - ak_2)[\vec{\rho}_u \times \vec{\rho}_v] \neq 0.$$

From the last formula, the first and second statements of the theorem follow. Furthermore, by Rodrigues's theorem,

$$(\vec{n}(P, a))_u = \vec{n}_u = -k_1\vec{\rho}_u, \qquad (\vec{n}(P, a))_v = \vec{n}_v = -k_2\vec{\rho}_v.$$

From this and from (2.43) we obtain

$$(\vec{n}(P, a))_u = \frac{-k_1}{1 - ak_1}\vec{r}_u, \qquad (\vec{n}(P, a))_v = \frac{-k_2}{1 - ak_2}\vec{r}_v. \qquad (2.44)$$

From (2.44) and Rodrigues's theorem we obtain

$$-k_1(P, a)\vec{r}_u = \frac{-k_1}{1 - ak_1}\vec{r}_u, \qquad -k_2(P, a)\vec{r}_v = \frac{-k_2}{1 - ak_2}\vec{r}_v. \qquad \square$$

2.5 The Third Fundamental Form of a Surface

On a regular surface Φ of class C^k ($k \geq 3$) one more fundamental form can be defined: the *third fundamental form*. Let $\vec{\lambda} = \lambda^1\vec{r}_u + \lambda^2\vec{r}_v$ be an arbitrary vector. Then suppose

$$III(\vec{\lambda}) = \left\langle \frac{d\vec{n}}{d\vec{\lambda}}, \frac{d\vec{n}}{d\vec{\lambda}} \right\rangle = \langle \lambda^1\vec{n}_u + \lambda^2\vec{n}_v, \lambda^1\vec{n}_u + \lambda^2\vec{n}_v \rangle = e(\lambda^1)^2 + 2f\lambda^1\lambda^2 + g(\lambda^2)^2,$$

where $e = \langle \vec{n}_u, \vec{n}_u \rangle$, $f = \langle \vec{n}_u, \vec{n}_v \rangle$, $g = \langle \vec{n}_v, \vec{n}_v \rangle$.

Note that $III(\vec{\lambda})$ induces a field of symmetric bilinear forms

$$III(\vec{\lambda}, \vec{\mu}) = e\lambda^1\mu^1 + f(\lambda^1\mu^2 + \lambda^2\mu^1) + g\lambda^2\mu^2.$$

It turns out that the three fundamental forms of a surface are linearly dependent.

Theorem 2.5.1. *At each point on a regular surface Φ of class C^k ($k \geq 3$) the following equality holds:*

$$K \cdot I(\vec{\lambda}) - 2H \cdot II(\vec{\lambda}) + III(\vec{\lambda}) = 0. \tag{2.45}$$

Recall that $K(P)$ is the Gaussian curvature of the surface at P, and $H(P)$ the mean curvature.

Proof. Let P be an arbitrary point on Φ. Introduce coordinates (u, v) in some neighborhood of this point such that the vectors $\vec{r}_u$ and $\vec{r}_v$ at P become parallel to the principal vectors. Then from Rodrigues's theorem and Theorem 2.4.3 we obtain, at the point P,

$$\langle \vec{r}_u, \vec{r}_v \rangle = 0, \tag{2.46}$$
$$\vec{n}_u = -k_1 \vec{r}_u, \quad \vec{n}_v = -k_2 \vec{r}_v. \tag{2.47}$$

From (2.46) and (2.47) follow

$$I(\vec{\lambda}) = E(\lambda^1)^2 + G(\lambda^2)^2,$$
$$II(\vec{\lambda}) = -\left\langle \vec{\lambda}, \frac{d\vec{n}}{d\vec{\lambda}} \right\rangle = k_1 E(\lambda^1)^2 + k_2 G(\lambda^2)^2, \tag{2.48}$$
$$III(\vec{\lambda}) = \left\langle \frac{d\vec{n}}{d\vec{\lambda}}, \frac{d\vec{n}}{d\vec{\lambda}} \right\rangle = k_1^2 E(\lambda^1)^2 + k_2^2 G(\lambda^2)^2.$$

From (2.48) we obtain

$$K \cdot I(\vec{\lambda}) - 2H \cdot II(\vec{\lambda}) + III(\vec{\lambda})$$
$$= k_1 k_2 \left[E(\lambda^1)^2 + G(\lambda^2)^2 \right]$$
$$\quad - (k_1 + k_2) \left[k_1 E(\lambda^1)^2 + k_2 G(\lambda^2)^2 + k_1^2 E(\lambda^1)^2 + k_2^2 G(\lambda^2)^2 \right]$$
$$= E(\lambda^1)^2 [k_1 k_2 - k_1(k_1 + k_2) + k_1^2]$$
$$\quad + G(\lambda^2)^2 [k_1 k_2 - k_2(k_1 + k_2) + k_2^2] = 0. \qquad \square$$

Remark 2.5.1. The equality (2.45) was proved for a special coordinate system, but since all characteristics in (2.45) are invariant, the equality is valid in any coordinate system. We write down (2.45) in more detailed form for an arbitrary coordinate system:

$$KE - 2HL + e = 0, \quad KF - 2HM + f = 0, \quad KG - 2HN + g = 0. \tag{2.49}$$

From (2.45) it follows that the third fundamental form of a surface does not itself play an essential role in the theory of surfaces. However, (2.45) or the equalities (2.49) may be useful for solving some interesting problems.

We shall give one example where the equalities (2.49) are used for proving Gauss's theorem about the area of a spherical image. Let us introduce two new notions.

Definition 2.5.1. Let D be some region on a regular surface Φ of class C^2. To each point on this surface we associate a point on the sphere $S^2(1)$ of unit radius by the following rule. Take a normal $\vec{n}(P)$ to Φ at a point P, and translate it by parallel displacement until the origin of the vector $\vec{n}(P)$ coincides with the center of the sphere $S^2(1)$. Then the endpoint of $\vec{n}(P)$ will coincide with some point $\varphi(P)$ on $S^2(1)$. The map $\varphi(P)\colon D \to S^2(1)$ thus constructed is called a *Gauss (spherical) map* of the region D on the surface Φ.

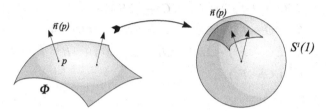

Figure 2.9. Gauss (spherical) map.

Definition 2.5.2. If D is some region on a surface Φ, then the real number $\omega(D) = \iint_D K\,dS$ is called the *integral curvature of* D. If D lies entirely in some coordinate neighborhood (u, v), then

$$\omega(D) = \iint_D K(u, v)\sqrt{E(u, v)G(u, v) - F^2(u, v)}\,du\,dv. \qquad (2.50)$$

Example 2.5.1. The Gauss (spherical) map of a sphere is a similarity. The spherical image of a cylindrical surface is a part of a *great circle* (i.e., an intersection with a plane containing the center of the sphere) on $S^2(1)$. The spherical image of a plane is a point.

Assume now that some region D on the surface Φ is in one-to-one correspondence with some region $D^* = \varphi(D)$ under the Gauss (spherical) map, and the Gaussian curvature of Φ has the same sign at each point of D. Then the following theorem holds.

Theorem 2.5.2 (Gauss theorem for a spherical map). *The modulus of the integral curvature of a region D is equal to the area of its spherical image*

$$|\omega(D)| = \iint_{D^*} dS_1, \qquad (2.51)$$

where dS_1 is an area element on a sphere.

Proof. Without loss of generality, assume that D lies inside a coordinate neighborhood; otherwise, divide it into parts, each of which lies in some coordinate neighborhood. Then prove the theorem separately for each such point and, in view of the additive nature of (2.51), deduce it for the entire region. Under this assumption the map φ is determined by the formula $\vec{r} = \vec{r}_1(u, v) = \vec{n}(u, v)$.

Consequently, for this parameterization the first fundamental form of a sphere coincides with the third fundamental form of Φ. But then

$$S(D^*) = \iint_{D^*} \sqrt{e(u,v)g(u,v) - f^2(u,v)} \, du \, dv. \qquad (2.52)$$

We find $\sqrt{eg - f^2}$, using (2.49), that

$$
\begin{aligned}
eg - f^2 &= (2HL - KE)(2HN - KG) - (2HM - KF)^2 \\
&= K^2(EG - F^2) - 2HK(LG - EN - 2MF) + 4H^2(LN - M^2) \\
&= K^2(EG - F^2) - 2HK \cdot 2H(EG - F^2) + 4H^2 \cdot K(EG - F^2) \\
&= K^2(EG - F^2).
\end{aligned}
$$

Consequently,

$$S(D^*) = \iint_D \sqrt{eg - f^2} \, du \, dv = \iint_D |K|\sqrt{EG - F^2} \, du \, dv = |\omega(D)|. \quad \square$$

Corollary 2.5.1. The ratio of the area of a spherical image of a region on a surface to the area of this region tends to the modulus of the Gaussian curvature at a given point P when a region ties up to this point and $K(P) \neq 0$.

Proof. If the Gaussian curvature at P differs from zero, then there is a neighborhood of P in which the Gaussian curvature has a fixed sign, and a Gauss (spherical) map of this region onto a sphere is single-valued. Then

$$\frac{S(D^*)}{S(D)} = \frac{\omega(D)}{S(D)} = \frac{|\iint_D K \, dS|}{S(D)} = \frac{|K(Q)| \iint_D dS}{\iint_D dS} = |K(Q)|,$$

where Q is some point in our region. If a region ties up to a point P, then $Q \to P$; thus $\lim \frac{S(D^*)}{S(D)} = |K(P)|$. $\qquad\square$

The Gauss theorem on the area of a spherical image has a perspective generalization. It is possible to add a clear geometrical sense to the formula (2.51): without the requirement of single-valuedness of a Gauss (spherical) map and constancy of the sign of Gaussian curvature, it can be written in the following form:

$$\omega(D) = \iint_D K \, dS = \iint_{D^*} dS_1.$$

To prove this formula and give it a geometric meaning, one needs a generalization of a notion of the area of a spherical image, which we omit for lack of space.

An application of the Gauss theorem on the area of a spherical image:

Let O be some point in $\mathbb{R}^3$. Denote by $P(O)$ the *set of all orientable planes in the space* $\mathbb{R}^3$ *passing through the point* O. The set $P(O)$ can be parameterized by the points of the unit sphere $S_1(O)$ with center at the point O, by corresponding to each point $Q \in S_1(O)$ a pair: a plane α through the point O that is orthogonal to the vector $\overrightarrow{OQ}$, and the vector $\overrightarrow{OQ}$ itself. Obviously, this map has an inverse.

Now let Φ be a regular surface of class C^k ($k \geq 2$). Denote by $P(\Phi, O)$ a *subset of all orientable planes of the set $P(O)$, for which the function $f_\alpha(P)$ has at least one degenerate critical point on Φ.*

Problem 2.5.1. Prove that the set $P(\Phi, O)$ is nowhere dense in the set $P(O)$ relative to the topology of the sphere $S_1(O)$.

Hint. Sketch of solution: It is not difficult to check that the point P_0 on Φ is a degenerate critical point of the function $f_\alpha(P)$ if and only if $T\Phi_P$ belongs to the set $P(\Phi, O)$. Now, using Gauss's theorem on spherical images, we see that the set $P(\Phi, O)$ has zero measure on the sphere $S_1(O)$. From this, the statement of the problem is deduced.

Problem 2.5.2. If the Gaussian curvature of a closed surface Φ is positive, then $|\omega(\Phi)| = 4\pi$.

2.6 Classes of Surfaces

Before starting with the material of this section, let us return to a discussion of the notion of a *surface*. The problem is that the definition given previously is rather coarse; it excludes the surfaces with points of self-intersection, but such surfaces often arise in natural geometric constructions, for instance, in the construction of parallel surfaces.

A surface in the sense of Definition 2.1.1 is an *embedded surface*; the term *two-dimensional manifold* embedded in $\mathbb{R}^3$ is also used. We introduce now a new class of surfaces: the *immersed surfaces*.

Definition 2.6.1. A set $\tilde{\Phi}$ is the *immersed surface* in $\mathbb{R}^3$ if there is an embedded surface Φ and a map $\varphi : \Phi \to \tilde{\Phi}$ that is a local diffeomorphism.

The difference between embedded and immersed surfaces is not essential when we study the local properties of a surface. Indeed, if a point P on an immersed surface $\tilde{\Phi}$ is a point of self-intersection, then we take the points P_1 and P_2 on Φ that are inverse images of P under the map φ. Select on the surface Φ the coordinate neighborhoods W_1 and W_2 of these points, each taken sufficiently small so that $W_1 \cap W_2 = \emptyset$ holds. Define $\tilde{W}_1 = \varphi(W_1)$ and $\tilde{W}_2 = \varphi(W_2)$.

Thus P has two coordinate neighborhoods on $\tilde{\Phi}$: a neighborhood $\tilde{W}_1$ on one "leaf" of a surface $\tilde{\Phi}$ and a neighborhood $\tilde{W}_2$ on a second "leaf," and local study of geometrical properties of $\tilde{\Phi}$ is reduced to local study of geometrical properties of first and second "leaves" separately. It turns out that at a given point $P \in \tilde{\Phi}$ we obtain two collections of geometric characteristics, which correspond to first and second "leaf," for instance, two values of the Gaussian curvature or two tangent planes. Thus, for local investigation of immersed surfaces, it is more convenient to imagine a point of self-intersection as two different points. Indeed, it may happen that there are not two, but three or more, "leaves" through a given point $P \in \Phi$.

During the study of the properties of surfaces *in the large* the difference between immersed and embedded surfaces can be essential. However, in most cases, a fact that holds for embedded surfaces is also true for immersed surfaces, but the proof in the latter case is often substantially more difficult.

We shall give one more notion that is useful for solving problems "in the large."

Definition 2.6.2. An *immersion* of a two-dimensional manifold (surface) Φ into $\mathbb{R}^3$ is called *proper* if the intersection of any compact subset of the space $\mathbb{R}^3$ with Φ is compact with respect to the intrinsic metric on Φ.

2.6.1 Surfaces of Revolution

Let a curve γ lie in the plane (x, z) and let its equation have the form $z = f(x)$. Assume that the function $f(x) \in C^2$ is monotonic — for simplicity, strongly monotonic. Denote by Φ the surface obtained by a rotation of γ about the axis OZ. The equation of this surface can be written in the following form:

$$z = f(\sqrt{x^2 + y^2}) = f(r), \quad r = \sqrt{x^2 + y^2}.$$

Using the formulas of Section 2.3, we find that

$$E = 1 + \left(\frac{x}{r} f'\right)^2, \quad G = 1 + \left(\frac{y}{r} f'\right)^2, \quad F = \frac{xy}{r^2}(f')^2, \quad EG - F^2 = 1 + (f')^2.$$

Since our surface is a surface of revolution, it is sufficient to find these geometrical characteristics at the points on some meridian of the surface, say $y = 0$. For points of this meridian, $F = 0$, $G = 1$, and $E = 1 + (f')^2$ hold. For the coefficients L, M, N of the second fundamental form we obtain

$$L = \frac{f''}{\sqrt{1 + (f')^2}}, \quad M = 0, \quad N = \frac{f'}{x\sqrt{1 + (f')^2}},$$

if a normal $\vec{n}(P)$ of a surface Φ is directed from the axis OZ of rotation. Hence, the Gaussian curvature is

$$K = \frac{f'' f'}{x[1 + (f')^2]^2}.$$

From visual observations it follows that K must be negative if the convexity of γ is directed along the axis OZ, and positive in the opposite case. In fact, if the convexity of γ is directed along the axis OZ, then $\frac{d^2 z}{dx^2} > 0$, but $\frac{d^2 z}{dx^2} = -\frac{f''}{(f')^3}$. Consequently, $-f'' f' > 0$ or $-f'' f' < 0$, and in the second case, $-f'' f' < 0$ or $-f'' f' > 0$; and since $x > 0$, our computations confirm the visual observations.

We now find the principal vectors and the principal curvatures. Hence the directions that are tangent to the meridian and parallels are the principal vectors. Moreover, from (2.25) we obtain

$$k_1 = \frac{L}{E} = \frac{f''}{[1+(f')^2]^{3/2}}, \quad k_2 = \frac{N}{G} = \frac{f'}{x\sqrt{1+(f')^2}}.$$

So, we see that k_1 is simply the curvature of γ, and the sign of k_1 is defined by the sign of f''.

We now clear up the geometrical sense of the principal curvature k_2. Draw a straight line through a point $P(x, f(x))$, which is orthogonal to the curve γ. The equation of this straight line is written as $(X - x) + f'(x)(Z - f(x)) = 0$, where X and Z are coordinates of a point on the straight line. We find the intersection of this straight line with the axis OZ. The point Q of intersection has coordinates $x = 0, z = (x + ff')/f'$. Denote by R the distance between P and Q and obtain

$$R = \sqrt{x^2 + \left(\frac{x+ff'}{f'} - f\right)^2} = \frac{x\sqrt{1+(f')^2}}{|f'|}.$$

Consequently, $|k_2| = 1/R$.

Remark 2.6.1. For a surface of revolution there often exists such a parameterization $\vec{r}(u, v)$ such that the coefficients of the first fundamental form are expressed as $E = 1, F = 0, G = G(u)$. In fact, if the parameter u is taken equal to $\sqrt{x^2 + y^2}$, and v equal to the rotation angle of the plane XOZ around the axis OZ, then

$$x = u\cos v, \quad y = u\sin v, \quad z = f(u)$$

are actually the parametric equations of our surface of revolution Φ. For this parameterization we have

$$\vec{r}_u = \cos v\vec{i} + \sin v\vec{j} + f'\vec{k}, \quad \vec{r}_v = -u\sin v\vec{i} + u\cos v\vec{j}.$$

Thus

$$E = 1 + (f')^2, \quad F = 0, \quad G = u^2.$$

We introduce a new parameter

$$\bar{u}(u) = \int_{u_0}^{u} \sqrt{1+(f')^2}\,du$$

and let $u = H(\bar{u})$ be the inverse function. Then $\vec{r}_{\bar{u}} = \vec{r}_u H'$, and hence we obtain

$$\bar{E} = (\vec{r}_{\bar{u}})^2 = \left[1 + (f')^2\right](H')^2 = \frac{1+(f')^2}{1+(f')^2} = 1, \quad F = 0, \quad G = H^2(\bar{u}),$$

which completes the proof of the remark.

Note finally that the parameter $\bar{u}$ has an obvious geometric sense: it is the arc length of the curve γ counting from a point $(u^0, f(u^0))$. In the case that γ cannot be defined by a monotonic function $f(x)$, it is more convenient to determine the equation of γ by a function of z: $x = \varphi(z)$. In this case we obtain the following equation of Φ:

$$\vec{r} = \vec{r}(u, v) = \varphi(u) \cos v\vec{i} + \varphi(u) \sin v\vec{j} + u\vec{k}.$$

We find the expression of the Gaussian curvature K of the surface Φ, using the geometric sense of the principal curvatures k_1 and k_2. We have

$$k_1 = -\frac{\varphi''}{[1 + (\varphi')^2]^{3/2}}$$

if the normal $\vec{n}(P)$ to Φ is directed along the axis of rotation. In order to find k_2, we must find the length of the subnormal of γ. The equation of the straight line through the point $P(\varphi(z), z)$ that is orthogonal to γ has the form

$$\varphi'(x - z) + Z - z = 0.$$

The coordinates of the point Q are $x = 0$, $Z = z + \varphi'\varphi$, and $R = \sqrt{\varphi^2 + (\varphi'\varphi)^2} = \varphi\sqrt{1 + (\varphi')^2}$. Consequently, $k_2 = \frac{1}{R} = \frac{1}{\varphi\sqrt{1+(\varphi')^2}}$ and $K = -\frac{\varphi''}{\varphi[1+(\varphi')^2]^2}$.

Problem 2.6.1. Find all surfaces of revolution with constant Gaussian curvature equal to K_0.

Solution. Obviously, this problem can be reduced to integration of the differential equation

$$K_0 = -\frac{\varphi''}{\varphi[1 + (\varphi')^2]^2}. \tag{2.53}$$

Multiplying (2.53) by $\varphi\varphi'$ and integrating, we obtain

$$-K_0\varphi^2 = -\frac{1}{1 + (\varphi')^2} + c. \tag{2.54}$$

To find the constant c it is necessary to know the initial conditions.

Consider the case of $K_0 > 0$. In this case the initial conditions are assumed to be

$$\varphi(0) = x_0, \quad \varphi'(0) = 0. \tag{2.55}$$

Then from (2.54) we obtain $c = 1 - K_0 x_0^2$. After this, we shall rewrite (2.54) in the form

$$(\varphi')^2 = \frac{-K_0(\varphi^2 - x_0^2)}{1 + K_0(\varphi^2 - x_0^2)}. \tag{2.56}$$

From (2.56) we see that its solution, the function φ, is an even function of the variable z, or in other words, the curve γ is symmetric with respect to the axis OX. Thus, consider (2.56) for $z < 0$, and then we can write it in the form

$$\varphi' = \frac{\sqrt{K_0(x_0^2 - \varphi^2)}}{\sqrt{1 + K_0(\varphi^2 - x_0^2)}}, \quad z < 0. \tag{2.57}$$

Generally, (2.57) is not integrable in elementary functions, but it is possible to present a parameterization of its solution as

$$x(\bar{u}) = x_0 \cos(\sqrt{K_0}\bar{u}), \quad z(\bar{u}) = \int_0^{\bar{u}} \sqrt{1 - x_0^2 K_0 \sin^2(\sqrt{K_0}t)} \, dt,$$

where the parameter $\bar{u}$, introduced above, is the arc length of the meridian. Consider three cases:

$$(1) \; x_0 = \frac{1}{\sqrt{K_0}}, \quad (2) \; x_0 > \frac{1}{\sqrt{K_0}}, \quad (3) \; x_0 < \frac{1}{\sqrt{K_0}}.$$

(1) In the first case the integral can be easily taken, and the solution has the form

$$x(\bar{u}) = x_0 \cos(\sqrt{K_0}\bar{u}), \quad z(\bar{u}) = x_0 \sin(\sqrt{K_0}\bar{u}).$$

Hence, γ is a semicircle, and the surface Φ is a sphere of radius $\frac{1}{\sqrt{K_0}}$. In the second and third cases we can make some qualitative observations. From the positivity of the expressions under the square roots in (2.57) we obtain a restriction on the solution φ of this equation in the form of the following inequalities:

$$K_0(x_0^2 - \varphi^2) > 0, \tag{2.58a}$$
$$1 - K_0(x_0^2 - \varphi^2) > 0. \tag{2.58b}$$

From (2.58a) we obtain $\varphi(z) \le x_0$, and from (2.58b) we obtain $K_0 \varphi^2 > K_0 x_0^2 - 1$. These inequalities allow us to see the differences between the second and the third cases on the one hand, and the first, on the other.

(2) $x_0 > \frac{1}{\sqrt{K_0}}$. In this case, $K_0 x_0^2 - 1 = a^2 > 0$ holds, and consequently, the function φ satisfies $\varphi(z) \ge |a|$, but $z'(\bar{u})$ at the point $\bar{u}_1 = \frac{1}{\sqrt{K_0}} \arcsin \frac{1}{x_0 \sqrt{K_0}}$ is zero. Also, we obtain that there is a real number

$$z_1 = \int_0^{\bar{u}_1} \sqrt{1 - x_0^2 K_0 \sin^2(\sqrt{K_0}t)} \, dt$$

such that the function $\varphi(z)$ is defined only in the interval $(-z_1, z_1)$, and in this interval it satisfies the inequality $0 < |a| < \varphi < x_0$; see Figure 2.10. Thus, we have obtained a nonclosed surface diffeomorphic to a cylinder.

(3) $x_0 < \frac{1}{\sqrt{K_0}}$. In this case, from the inequality (2.58a) we obtain the same estimate for the function $\varphi < x_0$. The second inequality (2.58b), in view of the condition $K_0 x_0^2 - 1 < 0$, shows us that there is some real number

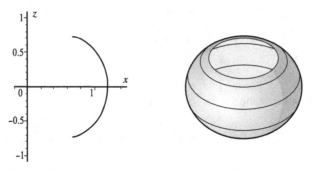

Figure 2.10. Case of $x_0 > \frac{1}{\sqrt{K_0}}$.

$$z_2 = \int_0^{\frac{\pi}{2\sqrt{K_0}}} \sqrt{1 - x_0^2 K_0 \sin^2(\sqrt{K_0}t)}\, dt$$

for which

$$\varphi(z_2) = 0, \quad \varphi'(z_2) = \frac{\sqrt{K_0}x_0}{\sqrt{1 - K_0 x_0^2}} > 0.$$

Hence, in this case the surface is homeomorphic to a sphere, but with two singular points $(0, z_2)$ and $(0, z_1)$; see Figure 2.11. The results obtained above are

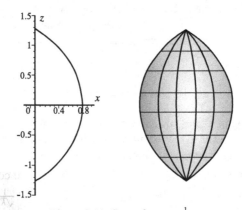

Figure 2.11. Case of $x_0 < \frac{1}{\sqrt{K_0}}$.

of course not fortuitous. We prove below that *any regular complete surface of constant positive Gaussian curvature is a sphere.*

Now consider the case of $K_0 < 0$. Assume $c = 1$ in (2.54). Then we obtain

$$K_0\varphi^2 = -\frac{(\varphi')^2}{1 + (\varphi')^2}.$$

For integration of this equation we pass to the parameterization of γ. Suppose $\varphi' = \tan t$. Then $K_0\varphi^2 = -\sin^2 t$, or

$$\varphi = \frac{1}{\sqrt{-K_0}} \sin t. \tag{2.59}$$

From (2.59) it follows that $dz = \frac{1}{\varphi'}\, dx = \cot t$. Consequently,

$$z = \frac{1}{\sqrt{-K_0}} \left(\cos t + \log \tan \frac{t}{2} \right) + c. \tag{2.60}$$

So, the required curve γ has the following parameterization:

$$x = \frac{1}{\sqrt{-K_0}} \sin t, \quad z = \frac{1}{\sqrt{-K_0}} \left(\cos t + \log \tan \frac{t}{2} \right) + c. \tag{2.61}$$

This curve is called a *tractrix*; see Figure 2.12. Its characteristic property is ex-

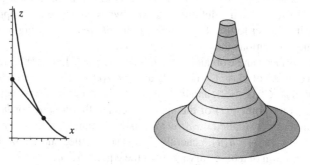

Figure 2.12. Tractrix and pseudosphere ($z \geq 0$).

pressed in the fact that *the length of a tangent line segment from a point of tangency to the z-axis is constant*. For $c = 0$ the length of this line segment is $\frac{1}{\sqrt{-K_0}}$. The obtained surface of revolution is called a *pseudosphere*. Its equations are

$$x = \frac{1}{\sqrt{-K_0}} \sin u \cos v, \quad y = \frac{1}{\sqrt{-K_0}} \sin u \sin v,$$

$$z = \frac{1}{\sqrt{-K_0}} \left(\cos u + \log \tan \frac{u}{2} \right).$$

The region of the parameters' values is determined by the inequalities

$$0 < u < \frac{\pi}{2}, \quad 0 \leq v < 2\pi.$$

The equality $z(u) = 0$ holds if and only if $u = \frac{\pi}{2}$, but then $\varphi(0) = \frac{1}{\sqrt{-K_0}}$, and $\lim_{z \to 0} \varphi'(z) = -\infty$. But if $\varphi(0) \neq \frac{1}{\sqrt{-K_0}}$, which corresponds to the value of the constant c in (2.58b) not equal to 1, then (2.54) is not integrable in elementary functions. Note that a pseudosphere is diffeomorphic to a cylinder. But there does not exist a complete surface in $\mathbb{R}^3$ with constant negative curvature that is diffeomorphic to a plane. This statement was proved by D. Hilbert at the end of nineteenth century.

Exercise 2.6.1. Find all minimal surfaces of revolution.

Another interesting subclass of surfaces of revolution is studied in Section 2.9.

2.6.2 Ruled and Developable Surfaces

Definition 2.6.3. A *ruled surface* is a one-parameter family of straight lines.

In general, the given definition of a ruled surface is not correct. It is not difficult to find the cases in which only some part of this set of points forms a surface. It is possible, of course, to assume in advance the regularity of a surface Φ and to call it *ruled* if it can be represented in the form of (or contains) a 1-parameter family of straight lines, for instance, a hyperboloid of one sheet or a hyperbolic paraboloid. However, such a definition is not always convenient for applications. Sometimes, it is better to use the *classical point of view*, according to which we always assume conditions that are necessary to us.

Any ruled surface can be obtained in the following way. Let $\gamma(t)$ be an arbitrary regular curve in $\mathbb{R}^3$ of class C^k ($k \geq 2$) and let $\vec{a}(t)$ be some vector field along $\gamma(t)$, also of class C^k ($k \geq 2$), and $\vec{a}(t) \neq 0$ for all $t \in (a, b)$. Construct a straight line through every point $\gamma(t)$ of the curve γ (called the *directrix* or *base curve*) in the direction of $\vec{a}(t)$ (called the *director curve*). A family of these straight lines (*rulings*), $u \rightarrow \vec{r}_1(u) + v\vec{a}(u)$, generally forms some ruled surface Φ. If $\vec{r}_1(u)$ is the parameterization of a base curve γ, then the equation $\vec{r}(u, v) = \vec{r}_1(u) + v\vec{a}(u)$ determines Φ. The answer to the question, what part of a set of points determined by this equation forms a regular surface depends on the vector functions $\vec{r}_1(u)$ and $\vec{a}(u)$. We shall not investigate this question in detail, but assume simply that on some region in the plane (u, v) this equation defines a regular surface. In every case, if $|\vec{a}(u)| = 1$ and $(\vec{a} \cdot \vec{a}' \cdot \vec{r}) \neq 0$ for all $u \in (a, b)$, the obtained ruled surface Φ is regular at each of its points.

Definition 2.6.4. A ruled surface with the condition $\vec{a}'(u) \neq 0$ is called *noncylindrical*. A noncylindrical ruled surface whose rulings are parallel to some fixed *directrix plane* is a *Catalan surface*. A Catalan surface is a *conoid* if all of its rulings intersect a constant straight line, called the *axis of the conoid*. A *conoid* is *right* if its axis is orthogonal to the directix plane.

Note that a right conoid is formed by a straight line that moves, guided on a fixed straight line orthogonal to it (the axis of conoid), and at the same time rotates about this straight line. If the velocity of rotation is proportional to the lifting velocity of the ruling, then the conoid is a *right helicoid*.

Example 2.6.1. The simplest conoid is the hyperbolic paraboloid. It is defined by moving a straight line that is parallel to a fixed plane and is guided by two fixed helices (two axes!). Conversely, every conoid that differs from a plane and has two axes is a hyperbolic paraboloid. The more complicated examples, *generalized*

Plücker's conoids (having $n \geq 2$ folds; see Figure 2.13[6] with $n = 2, 5$) are obtained by rotation of a ray about the axis OZ and with simultaneous oscillatory motion (with period $2\pi n$) along the segment $[-1, 1]$ of the axis:

$$\vec{r} = [0, 0, \sin(nu)] + v[\cos u, \sin u, 0] = [v \cos u, v \sin u, \sin(nu)].$$

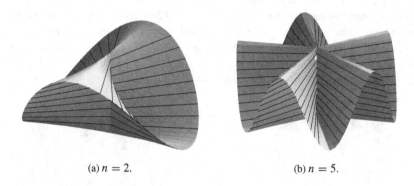

(a) $n = 2$.   (b) $n = 5$.

Figure 2.13. Generalized Plücker's conoids.

We now find an expression for the Gaussian curvature of a ruled surface

$$\vec{r}_u = \vec{r}_1{}' + \vec{a}'v, \quad \vec{r}_v = \vec{a}, \quad \vec{r}_{uu} = \vec{r}_1{}'' + \vec{a}''v, \quad \vec{r}_{uv} = \vec{a}', \quad \vec{r}_{vv} = 0.$$

From the last equality it follows that the coefficient N of the second fundamental form is zero. Thus the Gaussian curvature of Φ is expressed by the formula

$$K = -\frac{M^2}{EG - F^2}. \tag{2.62}$$

Formula (2.62) shows us that the Gaussian curvature of any ruled surface is non-positive, which of course does not surprise us, because at each point $P \in \Phi$ in the direction of the vector $\vec{a}$ the normal curvature is zero, and consequently, either the principal curvatures are of opposite signs or at least one of them is zero.

Consider in detail the case that the Gaussian curvature of a ruled surface is identically zero. We calculate the coefficient M to be

$$M = \frac{(\vec{r}_u \cdot \vec{r}_v \cdot \vec{r}_{uv})}{\sqrt{EG - F^2}} = \frac{((\vec{r}_1{}' + v\vec{a}') \cdot \vec{a} \cdot \vec{a}')}{\sqrt{EG - F^2}},$$

and consequently, the equality $K = 0$ means that

$$(\vec{r}_1{}'(u) \cdot \vec{a}(u) \cdot \vec{a}'(u)) = 0. \tag{2.63}$$

Equality (2.63) holds in the following obvious cases:

---

[6] For other examples of ruled surface modeling with Maple, see Figures 2.2, 2.14 and [Rov1].

1. $\vec{r}_1'(u) \equiv 0$. Hence the base curve γ is a point, and Φ is a *cone* (without the vertex).
2. $\vec{a}'(u) \equiv 0$. Hence $\vec{a}(u) = \text{const}$ and Φ is a *cylinder*.
3. $\vec{r}_1'(u) \times \vec{a}(u) = 0$. A surface Φ is generated by lines tangent to the base curve γ.

Example 2.6.2. Consider a *four-leafed rose* curve γ_1: $\rho = \cos(2\varphi)$, or in Cartesian coordinates of the $z = 1$ plane, $\vec{r}_1(u) = [\cos(2u)\sin(u), \cos(2u)\cos(u), 1]$. Then the *cylinder* with base curve $\gamma = \gamma_1$ and the z-axis is given by $\vec{r}(u, v) = \vec{r}_1(u) + v[0, 0, 1] = [\cos(2u)\cos(u), \cos(2u)\sin(u), 1]$. The *cone* with director curve $\vec{a} = \gamma_1$ and vertex $S(0, 0, 0)$ is given by $\vec{r}(u, v) = v\vec{r}_1(u)$. These surfaces, drawn with Maple, are shown in Figure 2.14.

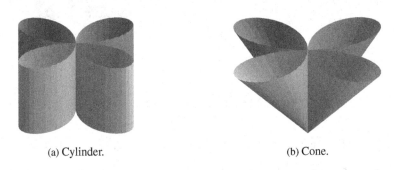

(a) Cylinder.   (b) Cone.

Figure 2.14. Cylinder and cone over a four-leafed rose.

Consider now the general case. From equality (2.63) it follows that

$$\vec{r}_1'(u) = \lambda_1(u)\vec{a}(u) + \lambda_2(u)\vec{a}'(u),$$

where λ_1 and λ_2 are some functions of the variable u. Note further that since $\vec{r}_1'(u) \times \vec{a}'(u) \neq 0$, the function $\lambda_2(u)$ is nonzero and

$$\vec{a} \times \vec{a}' \neq 0. \tag{2.64}$$

Take the curve $\tilde{\gamma}$ on a surface Φ defined by the equation $v = v(u)$ or $\vec{r}_2(u) = \vec{r}_1(u) + v(u)\vec{a}(u)$. We can select the function $v(u)$ such that $\vec{r}_2'(u) \times \vec{a} \equiv 0$. We obtain the equation for function $v = v(u)$,

$$(\vec{r}_1'(u) + v'(u)\vec{a}(u) + v(u)\vec{a}'(u)) \times \vec{a}(u) = 0,$$

or

$$(\lambda_1(u)\vec{a}(u) + \lambda_2(u)\vec{a}'(u) + v'(u)\vec{a}(u) + v(u)\vec{a}'(u)) \times \vec{a}(u) = 0,$$

or

$$(\lambda_2(u) + v(u)) \cdot (\vec{a}'(u) \times \vec{a}(u)) = 0$$

and from (2.64) it follows that $v(u) = -\lambda_2(u)$. So for the curve $\tilde{\gamma}$ and the vector field $\vec{a}(u)$ we obtain case 3. Consequently, for any ruled surface Φ with zero Gaussian curvature that is not a cone or a cylinder, there is a curve such that Φ is generated by the family of tangent lines to this curve. Surfaces with zero Gaussian curvature are called *developable*. In the next section we shall prove that *any developable surface is ruled*.

We now study the behavior of a tangent plane along generators of a developable surface Φ. Let $\vec{n} = \vec{n}(u, v)$ be a normal to Φ. Since the Gaussian curvature K is zero, a generator at each of its points goes along the main direction. Consequently, by Rodrigues's theorem,

$$\frac{\partial \vec{n}}{\partial v} = 0. \tag{2.65}$$

Equation (2.65) shows us that in fact the normal $\vec{n}(u, v)$ depends on u only; consequently, all tangent planes along a generator are parallel, and since they all contain the same straight line, they all coincide. This property of tangent planes along a generator is called the *stationarity of a tangent plane*.

Finally, we study in detail the structure of a developable surface in a neighborhood of a *base curve* $\tilde{\gamma}: \vec{r} = \vec{r}_2(u)$. Divide Φ onto two *semisurfaces* Φ_1 and Φ_2 by the inequalities $v \geq 0$ for Φ_1 and $v \leq 0$ for Φ_2. We now calculate the coefficients of the first fundamental form of the surface,

$$\vec{r}_u = (\vec{r}_2)_u + v\vec{a}'(u), \quad \vec{r}_v = \vec{a}(u). \tag{2.66}$$

If we assume u to be an arc length parameter of $\tilde{\gamma}$, and $\vec{a}(u) = \vec{r}_2'(u)$ a vector field, then from (2.66) it follows that

$$E = 1 + k^2 v^2, \quad F = G = 1, \tag{2.67}$$

where $k(u)$ is the curvature of $\tilde{\gamma}$. From (2.67) one can see that at the points $(u, v) \in \Phi_1$ and $(u, -v) \in \Phi_2$, the coefficients E, F, and G coincide. Consequently, the semisurfaces Φ_1 and Φ_2 are isometric. Thus Φ is the union of two isometric semisurfaces Φ_1 and Φ_2 that are *"glued" to each other* along the base curve. Thus, we see that the surface Φ is not regular at the points of the base curve $\tilde{\gamma}$. At its other points, Φ is regular if the curvature $k(u)$ of $\tilde{\gamma}$ is nonzero at each point, because $|\vec{r}_u \times \vec{r}_v|^2 = EG - F^2 = k^2 v^2$.

The semisurfaces Φ_1 and Φ_2 often intersect with each other. Here is the simplest example: Let γ be a circle in $\mathbb{R}^2$, and suppose the field $\vec{a}(t)$ is a tangent vector field to γ. Then Φ_1 and Φ_2 obviously coincide, and Φ consists of two copies of the exterior of a circle, "glued" along a circle.

We now explain why a ruled surface Φ with zero Gaussian curvature is called a developable surface. Let $P_0(u_0, v_0)$ be a regular point of Φ. Take on Φ a neighborhood U of P_0 defined by the inequalities $|u - u_0| < \varepsilon$, $|v - v_0| < \varepsilon$. Let ε be sufficiently small such that the curvature $k(u)$ of the base curve differs from zero if $|u - u_0| < \varepsilon$, and $v \neq 0$ if $|v - v_0| < \varepsilon$. Take a curve $\bar{\gamma} : \vec{r} = \bar{\rho}(u)$ on some plane α such that its curvature $\bar{k}(u)$ coincides with $k(u)$ for $|u - u_0| < \varepsilon$ and $\vec{a}(u)$ is the unit vector field tangent to $\bar{\gamma}$. Then the region $\bar{U}$ of a plane α determined by the condition $\vec{r} = \vec{\rho}(u) + v\vec{a}(u)$ for $u, v \in U$ is isometric to the region U on Φ, as is seen from (2.67). So, we can say that a neighborhood U of P_0 is developed (unrolled) onto a plane.

Finally, note that if a complete regular surface Φ is developable, then Φ is a cylinder (in particular, a plane).

2.6.3 Convex Surfaces

Recall that a *region* $D \subset \mathbb{R}^3$ is *convex* if together with each pair of its points P and Q it contains the line segment PQ connecting these points. The boundary ∂D of a convex region D is actually a surface (generally, a continuous surface), and it is called a *convex surface*.

Theorem 2.6.1. *If a convex region D contains a straight line, then the surface $\Phi = \partial D$ is homeomorphic to a cylinder; if D contains a ray but does not contain any straight line, then $\Phi = \partial D$ is homeomorphic to a plane; if D does not contain any ray (D is compact), then the surface $\Phi = \partial D$ is homeomorphic to a sphere.*

Proof. The statements of Theorem 2.6.1 are straightforward, and we shall prove only the first of them. Let a be a straight line that lies entirely in D, and let P be an arbitrary point on Φ. Take two points P_1 and P_2 on the straight line a. In view of the convexity of D, the line segments PP_1 and PP_2 belong to D. Let points P_1 and P_2 tend to infinity along the straight line a, and let the length of the line segment $P_1 P_2$ also tend to infinity. Then the line segments PP_1 and PP_2 tend to some rays a_1 and a_2 lying on the straight line $a(P)$ that is parallel to a. Consequently, through each point $P \in \Phi$ there passes a straight line $a(P)$ that is parallel to a and lies entirely on Φ. Bring the plane α through some point $Q \in \Phi$ and orthogonally to a. Since $\alpha \cap D$ is a convex region on the plane α, the intersection $\alpha \cap \Phi$ is a convex curve γ. Let C be a cylinder whose directrix curve is γ and whose rulings are parallel to the straight line a. Since C is contained in Φ and is an open and closed set, they coincide, which completes the proof. $\square$

The following obvious properties of convex surfaces are formulated as exercises.

Exercise 2.6.2. A regular surface Φ of class C^k ($k \geq 1$) is convex if and only if it lies entirely on one side of every one of its tangent planes.

Exercise 2.6.3. The Gaussian curvature of a regular convex surface of class C^k ($k \geq 2$) is nonnegative at every point.

Exercise 2.6.4. The integral curvature (see Section 2.5) of a closed regular convex surface of class C^2 is 4π.

Hint. Use Gauss's Theorem 2.5.2 (about spherical images).

Exercise 2.6.5. The exact upper bound of the integral curvature of any open, regular, convex surface of class C^2 is not greater than 2π.

Hint. Use Gauss's Theorem 2.5.2 (about spherical images).

Exercise 2.6.6. The exact lower bound of the Gaussian curvature $K(P)$ of an open, regular, convex surface of class C^2 is zero.

Hint. Use the statement of the previous exercise.

Exercise 2.6.7. If the Gaussian curvature $K(P)$ of a regular convex surface of class C^2 is not smaller than a positive real number a, then the surface is closed (compact).

Exercise 2.6.8. If a convex surface Φ is neither a cylinder nor a plane, then there is a nondegenerate right circular cone containing Φ.

2.6.4 Problems: Curvature of Convex Surfaces

Now we formulate some conditions under which a regular surface of class C^2 has points of positive Gaussian curvature.

Problem 2.6.2. On each closed (compact) regular surface Φ of class C^2 there is a point at which the Gaussian curvature is positive.

Solution. Let $S(O, R)$ be a sphere with center at some point O and radius R so large that the entire surface Φ lies in this sphere. Decrease the radius of this sphere until the sphere and Φ touch each other for the first time, at which time the radius of the sphere will be $R_1 < R$. Denote by P a point that belongs to $\Phi \cap S(O, R_1)$. The tangent plane $T\Phi_P$ at P is also a tangent plane to the sphere $S(O, R_1)$. Direct a normal $\vec{n}(P)$ inside of $S(O, R_1)$. The normal curvatures of $S(O, R_1)$ at this point are equal to $1/R_1$, and the normal curvatures of Φ are not smaller than $1/R_1$, because Φ lies entirely inside the sphere $S(O, R_1)$. Consequently, the Gaussian curvature $K(P)$ of Φ at P is not smaller than $1/(R_1)^2$.   □

An analogous criterion for complete, noncompact (open) surfaces can be formulated as follows:

Problem 2.6.3. If an open, regular surface Φ of class C^2 lies entirely inside a convex circular cone T, then there is a point on Φ at which the Gaussian curvature is positive.

Solution. Write down the equation of the cone T in the form $x^2 + y^2 - a^2 z^2 = 0$, and define a region containing the surface Φ by the inequalities

$$x^2 + y^2 - a^2 z^2 < 0, \quad z > 0.$$

Take another cone T_1, defined by the equation

$$x^2 + y^2 - b^2(z+c)^2 = 0 \quad \text{(where} \quad b > a, c > 0). \tag{2.68}$$

Let H be one (upper) sheet of the hyperboloid of revolution of two sheets given by the equation

$$\frac{4b^2}{c^2}(x^2 + y^2) - \frac{4}{c^2}(z+c)^2 = -1, \quad z \geq -\frac{c}{2}.$$

This convex surface H lies entirely between the cones T and T_1, and the cone T_1 is asymptotic to H. Let

$$\rho_0 = \inf_{P \in D, Q \in H} |PQ|.$$

From the condition $b > a$ in (2.68), follows that ρ_0 exists and is finite, and that there exist points P_0 and Q_0 on the surfaces Φ and H such that $\rho_0 = |P_0 Q_0|$. Now displace H parallel to itself by the vector $\overrightarrow{P_0 Q_0}$. Denote by H_0 the resulting surface. The surfaces H_0 and Φ have a common tangent plane at P_0, and Φ lies entirely inside the convex region bounded by H_0. Thus, as in the previous problem, the Gaussian curvature of the surface Φ at P_0 is not smaller than the Gaussian curvature of the surface H_0 at Q_0, which is positive. □

Remark 2.6.2. If a convex surface Φ has a point at which the Gaussian curvature is positive, then there is a right circular cone T containing Φ (see Exercise 2.6.8).

Thus from Problem 2.6.3 we obtain the following corollary.

Corollary 2.6.1. If a complete, open regular surface Φ lies entirely inside a convex region bounded by a convex surface Φ_1 whose Gaussian curvature is positive at least at one point, then there is a point on Φ at which the Gaussian curvature is positive.

We now solve Hadamard's problem.

Problem 2.6.4 (Hadamard). If the Gaussian curvature at each point of a closed regular surface Φ of class C^2 is positive, then Φ is convex.

Hadamard's problem is a particular case of Problem 2.6.5 given below. We shall give another solution of this problem, different from but no less illuminating than the solution of Problem 2.6.5.

Solution. The surface Φ bounds some three-dimensional region, which will be denoted by D. Take a point P on Φ, and direct a normal $\vec{n}(P)$ to Φ at this point P inside of D, and continue this direction of the normal by continuity to each point on Φ. Define the sign of the normal curvature at each point of Φ by the direction of the normal $\vec{n}(P)$. In view of the statement of Problem 2.6.2, there is a point Q on Φ at which under the above definition of the sign of a normal curvature all normal curvatures are positive. By the conditions of our problem, the Gaussian curvature at all points $P \in \Phi$ is positive. It follows that the normal curvatures of Φ are positive at all of its points. Now let Q be some point in int D. Denote by

Γ_Q the set of points int D that can be connected with Q by a line segment entirely contained in int D. Obviously, Γ_Q is an open set in int D.

We prove that Γ_Q is closed in int D. Let $Q_0 \in \text{int } D$ be a point such that there is a sequence of points $Q_n \in \Gamma_Q$ tending to Q_0. Assume that $Q_0 \notin \Gamma_Q$. That means that the line segment QQ_0 touches Φ at some point P that lies inside of the line segment QQ_0 ($P \in QQ_0$, $P \neq Q$, $P \neq Q_0$). Then $\overrightarrow{PQ_0}$ belongs to the tangent plane $T\Phi_P$, and the normal curvature of Φ at the point P and in the direction of the vector $\overrightarrow{PQ_0}$ is nonpositive, which is a contradiction. $\square$

Problem 2.6.5. Let Φ be a closed, regular surface of class C^2 immersed in $\mathbb{R}^3$. Then if the Gaussian curvature of the surface Φ at each of its points is nonnegative, then Φ is convex, and consequently, an embedded surface.

Solution. Assume the opposite and lead this assumption to a contradiction. If Φ is a nonconvex surface, then there is a plane α such that the set $\alpha \cap \Phi$ is not connected, because in the opposite case, Φ will lie on one side of each of its tangent planes and will be convex. Take a point Q on the plane α and a unit vector $\vec{e}$ orthogonal to the plane α. Define a function $f_\alpha(P)$ on Φ, assuming it to be equal to $(\overrightarrow{QP}, \vec{e})$. Note that in view of Problem 2.5.1, we can assume that $f_\alpha(P)$ has no degenerate critical points on the surface Φ. Let P_1 and P_2 be two points in the set $\Phi \cap \alpha$ from its different connected components. Without loss of generality, we may suppose that there is a continuous curve $\sigma(t)$ ($0 \leq t \leq 1$) on Φ connecting P_1 and P_2 and belonging to the region defined by the inequality $f_\alpha(P) \geq 0$. Let $\Gamma(P_1 P_2)$ be the class of all continuous curves σ on Φ with endpoints P_1 and P_2 that lie in the region $f_\alpha(P) \geq 0$. Take a point P_σ on each curve $\sigma \in \Gamma(P_1 P_2)$ at which the function $f_\alpha(\sigma(t))$ for $0 \leq t \leq 1$ reaches its maximum. Define a real number a_0 by the equality

$$a_0 = \inf_{\sigma \in \Gamma(P_1 P_2)} f_\alpha(P_\sigma).$$

Obviously, $a_0 > 0$. In view of the compactness of Φ and the definition of a_0, there is a point P_0 on Φ such that $f_\alpha(P_0) = Q_0$; and there exist points $Q_1 \neq P_0$ and $Q_2 \neq P_0$ in any neighborhood of P_0 on the surface Φ such that

$$f_\alpha(Q_1) \leq a_0, \tag{2.69}$$

$$f_\alpha(Q_2) \geq a_0. \tag{2.70}$$

In view of the definition of a_0 and of the point P_0, a plane $T\Phi_{P_0}$ is parallel to the plane α. Introduce a rectangular Cartesian coordinate system x, y, z with the origin at P_0; direct the axis OZ parallel to the vector $\vec{e}$, and so the axes OX and OY are located in the plane $T\Phi_{P_0}$. The equation of Φ in some neighborhood of P_0 can be written explicitly as

$$z = h(x, y).$$

The function $h(x, y)$ satisfies the equations

$$h(0, 0) = 0, \quad \frac{\partial h}{\partial x}(0, 0) = \frac{\partial h}{\partial y}(0, 0) = 0;$$

moreover, the axes OX and OY can be chosen so that

$$\frac{\partial^2 h}{\partial x \partial y}(0, 0) = 0.$$

We introduce the notation

$$a = \frac{1}{2}\frac{\partial^2 h}{\partial x^2}(0, 0), \quad b = \frac{1}{2}\frac{\partial^2 h}{\partial y^2}(0, 0).$$

Write Taylor's formula for the function $h(x, y)$:

$$h(x, y) = ax^2 + by^2 + \bar{o}(x^2 + y^2). \tag{2.71}$$

Since the functions $h(x, y)$ and $f_\alpha(P)$ in a neighborhood of the point P_0 are related by the equality $f_\alpha(x, y) = h(x, y) + a_0$, and since $f_\alpha(0)$ has no degenerate critical points on Φ, P_0 is a degenerate critical point of the function $h(x, y)$. Thus $a \cdot b \neq 0$ holds. If we assume that $a \cdot b > 0$, then in view of (2.71), this leads to a contradiction either with the inequality (2.69) or with the inequality (2.70). Consequently, $a \cdot b < 0$. But then the Gaussian curvature K of Φ, equal to $4a \cdot b$, would be negative, which contradicts the conditions of the problem. The problem is solved. □

Now we study open (i.e., complete noncompact) surfaces.

Problem 2.6.6. If the Gaussian curvature of an open regular surface Φ of class C^2 is everywhere positive, then Φ is convex.

Hint. Repeat the considerations and constructions of Problem 2.6.4.

An analogue of Problem 2.6.5 appears to be more complicated. The following statement holds.

Problem 2.6.7. Let Φ be an open regular surface of class C^2 immersed in $\mathbb{R}^3$. Then if the Gaussian curvature of Φ at each of its points is nonnegative and there is a point of positive Gaussian curvature, then Φ is convex and consequently, an embedded surface.

The solution of this problem is difficult and long. It can be found in [Pog]. We now give some tests that allow us to estimate from above the principal curvatures of a surface at some points of this surface.

Problem 2.6.8. If a region D bounded by a convex surface Φ of class C^2 contains a ball $C(R)$ of radius R, then there is a point P on Φ at which all normal curvatures are not greater than $1/R$.

Solution. Increase the radius of the ball $C(R)$ while its center is fixed until it (the ball $C(R_1)$, $R_1 > R$) touches the surface Φ for the first time at some point P_0. So we have obtained that Φ touches the ball $C(R_1)$ at P_0 and lies outside of the ball $C(R_1)$. From this it follows that the absolute value of any normal curvature is not greater than $1/R_1$, and consequently, than $1/R$. □

Remark 2.6.3. The convexity assumption for the surface Φ is not essential when the sign of the normal curvature of Φ is defined by the direction of its inner normal.

This test is essentially strengthened in the following problem.

Problem 2.6.9. Let Φ be a convex, regular surface of class C^2 bounded by a convex region D. Then, if one can place a circle with radius R entirely inside of D, there is a point on Φ at which all the normal curvatures are not greater than $1/R$. (The sign of the normal curvature is determined by the direction of the inner normals.)

Solution. Let K_R be a disk of radius R that lies entirely in D, and let α be the plane containing this disk. Denote by C the right circular cylinder whose generatrix is the boundary of the disk K_R, and the rulings are straight lines orthogonal to the plane α. The following two cases are possible:

(1) the cylinder C does not intersect the surface Φ,

(2) the cylinder C cuts out from the surface Φ at least one surface Φ_1.

In the first case, as follows from Theorem 2.6.1, Φ itself is a cylinder C, and the statement of the problem becomes obvious.

Consider the second case. Let $\tilde{D}$ be the region bounded by the cylinder C and Φ_1. Let $S(R - \varepsilon)$ be the sphere of radius $R - \varepsilon$ ($0 < \varepsilon < R$) with center on the axis of the cylinder C that does not intersect Φ_1. Move the sphere $S(R - \varepsilon)$ in the direction toward Φ_1 until the first tangency of $S(R - \varepsilon)$ and Φ_1 appears, and let P be a point of tangency. The surface Φ_1, as follows from our construction, lies entirely on one side of $S(R - \varepsilon)$ and touches it at P, an interior point of Φ_1. Thus all normal curvatures of Φ_1 at P are not greater than $\frac{1}{R-\varepsilon}$. Since ε may be chosen as close to zero as possible, the statement of the problem is proved. □

There are many books and surveys devoted to convex surfaces and bodies without any smoothness assumptions; see [Ku1], [Bus], [Sto], [Pog], [Hop].

2.6.5 Saddle Surfaces

Definition 2.6.5. A regular surface Φ of class C^2 is called a *saddle surface* if the Gaussian curvature at each point of Φ is nonpositive.

Note that the class of saddle surfaces contains the well-studied class of *minimal surfaces*, i.e., surfaces whose mean curvature H is identically zero,[7]

---

[7] See, for example, [OG].

Generally speaking, there is another definition of a saddle surface that is also applicable for a continuous surface and coincides with our definition in the regular case.

If the Gaussian curvature of Φ is negative at some point $P \in \Phi$, then the tangent plane $T\Phi_P$ intersects Φ, and the surface lies on both sides of $T\Phi_P$. The surface Φ has a *saddle*-type shape in a neighborhood of such a point, which is the reason for the name of these surfaces. One of the features of a complete saddle surface is its unboundness in $\mathbb{R}^3$. So for instance, from Problem 2.6.2 it follows that there does not exist a closed saddle surface in $\mathbb{R}^3$, and from Problem 2.6.3 (or its Corollary 2.6.1) it follows that *one cannot place a saddle surface entirely in any strictly convex region.* But a saddle surface can be entirely placed between two parallel planes, for instance, a circular cylinder or, a more interesting example, the surface obtained by rotation of the curve $x = \frac{z^2}{1-z^2}$ ($|z| < 1$) about the axis OZ. The Gaussian curvature of this surface is negative at each point. But this procedure is impossible for saddle surfaces homeomorphic to a plane.

Theorem 2.6.2 (S.N. Bernstein). *If a saddle surface Φ defined by the equation*

$$z = f(x, y) \quad (-\infty < x, \, y < \infty)$$

has points of negative Gaussian curvature, then

$$\sup_{x,y} |f(x, y)| = \infty.$$

The proof of this theorem is very difficult and cannot be given here. We now study the behavior of the Gaussian curvature $K(P)$ and the principal curvatures $k_1(P)$ and $k_2(P)$ of saddle surfaces.

The strongest result concerning the behavior of the Gaussian curvature on a saddle surface was obtained by N.V. Efimov.

Theorem 2.6.3 (N.V. Efimov). *The least upper bound of the Gaussian curvature on a complete saddle surface in $\mathbb{R}^3$ is zero.*

This theorem of N.V. Efimov is very deep and difficult to prove. It is an essential generalization of the well-known theorem of Hilbert on the *nonexistence in $\mathbb{R}^3$ of a complete regular surface with constant negative Gaussian curvature.* In this book we cannot give its proof. We restrict ourselves to only the proof of Hilbert's theorem given at the end of Section 2.8.

The statements of N.V. Efimov's theorem and Exercise 2.6.6 can be formulated in the form of a single theorem.

Theorem 2.6.4. *The exact lower bound of the absolute value of the Gaussian curvature of a complete open regular surface in $\mathbb{R}^3$ is zero.*

Let us now study the behavior of the principal curvatures $k_1(P)$ and $k_2(P)$ at infinity for any complete saddle surface.

Problem 2.6.10. Prove that if on a complete saddle surface $\Phi \subset \mathbb{R}^3$,

$$\inf_{P \in \Phi} |k_1(P)| + \inf_{P \in \Phi} |k_2(P)| = c > 0$$

holds, then Φ is a cylinder, i.e., $k_1(P) \equiv 0$ and $k_2(P) = \text{const.}$

Solution. Assume for definiteness that $k_1(P) \leq 0$ and $k_2(P) \geq 0$ and that $\inf_{P \in \Phi} |k_2(P)| = c_1 > 0$. Assume that the Gaussian curvature $K(P)$ is not identically zero, and show that this leads to a contradiction. Take a real number R, satisfying the inequality

$$R > \frac{1}{c_1}. \tag{2.72}$$

Then the Gaussian curvature of a parallel surface $\Phi(R)$ is nonnegative. Indeed, by Theorem 2.4.6 we have

$$k_1(P, R) = \frac{k_1(P)}{1 - Rk_1(P)}, \quad k_2(P, R) = \frac{k_2(P)}{1 - Rk_2(P)}.$$

From this, we obtain

$$K(P, R) = \frac{K(P)}{(1 - Rk_1(P))(1 - Rk_2(P))}.$$

The numerator and denominator of this fraction are negative. Indeed, $K(P) \leq 0$ by the conditions of the problem, $1 - Rk_1(P) > 0$ because of $k_1(P) \leq 0$, and $1 - Rk_2(P) < 0$ in view of inequality (2.72). From the nonnegativity of the Gaussian curvature of $\Phi(R)$, it follows (see Exercise 2.6.8) that $\Phi(R)$ can be placed entirely inside of a strictly convex cone C. Take another strongly convex cone C_1 such that it contains C and the distance from any point of C to C_1 is greater than R. Then a saddle surface Φ lies entirely in the cone C, which is impossible (see Problem 2.6.3). Thus, the Gaussian curvature of Φ is identically zero.   □

In 1966, J. Milnor enunciated a conjecture, which for saddle surfaces implies that for statement 2.6.10 to hold, it suffices that:

$$\inf_{P \in \Phi} (|k_1(P)| + |k_2(P)|) \neq 0.$$

This assertion is not yet been proved.

We solve one more problem, in which saddle and convex surfaces are closely related.

Problem 2.6.11. Let Φ be a regular convex surface of class C^k $(k \geq 2)$, and $k_1(P)$ and $k_2(P)$ the principal curvatures of Φ at a point P. Suppose that $0 \leq k_1(P) \leq k_2(P)$. In this case, if $\sup_{P \in \Phi} k_1(P) < \inf_{P \in \Phi} k_2(P)$, then Φ is a cylinder, and consequently, $k_1 \equiv 0$, but $k_2 = c_0 > 0$. The sign of the normal curvature is determined by the direction of the inner normal.

Solution. We assume that Φ is not a cylinder, and show this leads to a contradiction. The surface Φ cannot be homeomorphic to a sphere, because on any such surface there is an *umbilic* at which $k_1(P) = k_2(P)$, in contradiction to the condition of the problem. Thus we need only study the case that Φ is homeomorphic to a plane; in this case we may suppose that $\sup_{P \in \Phi} k_1(P) = c_1 > 0$. Define $c_2 = \inf_{P \in \Phi} k_2(P)$, and denote by R a real number satisfying the inequality

$$\frac{1}{c_2} < R < \frac{1}{c_1}. \tag{2.73}$$

Take parallel surfaces $\Phi(R)$ and $\Phi(-R)$. The surface $\Phi(-R)$ is convex and contains $\Phi(R)$ entirely. Prove that $\Phi(R)$ is regular and a saddle surface. The regularity of $\Phi(R)$ follows from Theorem 2.4.6. We now calculate the principal curvatures of $\Phi(R)$ at a point $\varphi(P)$. (We use the notation introduced in Section 2.4.9). By Theorem 2.4.6,

$$k_1(P, R) = \frac{k_1(P)}{1 - Rk_1(P)}, \quad k_2(P, R) = \frac{k_2(P)}{1 - Rk_2(P)}.$$

Thus the Gaussian curvature $K(P, R)$ of $\Phi(R)$ is expressed by the formula

$$K(P, R) = \frac{K(P)}{(1 - Rk_1(P))(1 - Rk_2(P))}. \tag{2.74}$$

The numerator of this fraction is nonnegative, and the denominator is negative. In fact, in view of (2.73),

$$1 - Rk_1(P) \geq 1 - c_1 R > 0, \tag{2.75}$$

but

$$1 - Rk_2(P) \leq 1 - c_2 R < 0. \tag{2.76}$$

Thus, from (2.74) – (2.76) follows that

$$K(P, R) \leq 0.$$

So a saddle surface $\Phi(R)$ lies entirely inside of the convex surface $\Phi(-R)$, which is impossible. ☐

2.7 Some Classes of Curves on a Surface

2.7.1 Lines of Curvature

Definition 2.7.1. A smooth curve γ on a regular surface Φ of class C^2 is called a *line of curvature* if the tangent vector of γ is a principal vector of Φ at all points of γ.

Recall the equation determining the principal vectors

$$\begin{vmatrix} -(\lambda^2)^2 & \lambda^1\lambda^2 & -(\lambda^1)^2 \\ E & F & G \\ L & M & N \end{vmatrix} = 0.$$

If $u = u(t)$, $v = v(t)$ are equations of some curve, then its tangent vector has in a local base the coordinates $u'(t)$ and $v'(t)$. Thus the problem of existence of the lines of curvature is reduced to the problem of the existence of a solution to the following differential equation:

$$(LF - EM)\left(u'\right)^2 - (EN - LG)u'v' + (MG - NF)\left(v'\right)^2 = 0.$$

From Lemma 2.1.1 it follows that through each nonumbilical point of the surface Φ it is possible to pass a line of curvature and then to extend it until we reach an umbilic.

From the same lemma it follows that in a neighborhood of any nonumbilical point one can construct a coordinate system such that the lines of curvature will become the coordinate curves. The characteristic indication of such a coordinate system is the fulfilment of the following equalities: $M = F = 0$. In fact, if the coordinate curves are the lines of curvature, then $F = 0$ in view of their orthogonality, and $M = 0$, since the principal vectors are conjugate with respect to the second fundamental form.

We now consider some geometric properties of the lines of curvature.

Theorem 2.7.1. *A ruled surface C formed by the family of normal straight lines to Φ along the lines of curvature has nonzero Gaussian curvature.*

Proof. Let $\vec{r} = \vec{r}(t)$ be the equation of a line of curvature γ, and $\vec{n}(t)$ the directions of the normals to Φ along γ. Then the equation of the surface C can be written in the form

$$\vec{r} = \vec{r}(u, v) = \vec{r}(u) + v\vec{n}(u).$$

By Rodrigues's theorem, $\vec{n}'(u) = -k\vec{r}'(u)$, and consequently, see (2.64), the Gaussian curvature of the surface Φ is zero.   □

Theorem 2.7.2. *If two surfaces Φ_1 and Φ_2 intersect with a constant angle and the curve of intersection is a line of curvature on one of them, then it is also a line of curvature on the other surface.*

Proof. Let $\vec{r} = \vec{r}(t)$ be the equation of the curve of intersection of the surfaces Φ_1 and Φ_2. Denote by $\vec{n}_1$ and $\vec{n}_2$ the unit normals to Φ_1 and Φ_2 along γ, respectively. From the conditions of the theorem we obtain the equations

$$\langle \vec{n}_1(t), \vec{n}_2(t) \rangle = \text{const}, \tag{2.77}$$

$$\frac{d\vec{n}_1}{dt} = -k(t)\vec{r}'(t), \tag{2.78}$$

$$\langle \vec{r}'(t), \vec{n}_1 \rangle = \langle \vec{r}'(t), \vec{n}_2 \rangle = 0. \tag{2.79}$$

From (2.77) follows

$$\langle d\vec{n}_1/dt, \vec{n}_2 \rangle + \langle \vec{n}_1, d\vec{n}_2/dt \rangle = 0, \tag{2.80}$$

and from (2.78) and (2.80) follows

$$\langle \vec{n}_1, d\vec{n}_2/dt \rangle = 0. \tag{2.81}$$

Moreover, we have

$$\langle \vec{n}_2, d\vec{n}_2/dt \rangle = 0. \tag{2.82}$$

Thus, from the equalities (2.81) and (2.82) we see that the vector $\frac{d\vec{n}_2}{dt}$ is orthogonal to $\vec{n}_1$ and $\vec{n}_2$, but then from the equality (2.79) follows the collinearity of $\vec{r}'$ and $\frac{d\vec{n}_2}{dt}$, i.e., $\frac{d\vec{n}_2}{dt} = \alpha \vec{r}$, and now the theorem follows from Rodrigues's theorem.   □

Corollary 2.7.1. If a plane or a sphere intersects some surface with a constant angle, then the curve of intersection is the line of curvature.

This statement follows from the fact that *any curve on a plane and on a sphere is the line of curvature*. From this it follows that *all parallels and meridians on a surface of revolution are its lines of curvature*.

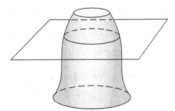

Figure 2.15. Curvature lines (parallels) on a surface of revolution.

Finally, we study the properties of the lines of curvature on a surface Φ with zero Gaussian curvature. Let $\gamma(t)$ be the line of curvature that passes along the principal vector corresponding to zero principal curvature. Then by Rodrigues's theorem, $\frac{d\vec{n}}{dt} = 0$, and consequently, the field of normals to Φ along $\gamma(t)$ is constant: $\vec{n}(t) = \vec{n}(0) = \vec{c}$. But then $\langle \vec{r}'(t), \vec{n}_0 \rangle = 0$ and $\langle \vec{r}(t) - \vec{r}(0), \vec{n}_0 \rangle = $ const. The last equality means that γ is a plane curve. We prove that in the case $H(P) \neq 0$ a curve γ is a line segment. Introduce in a neighborhood of γ a coordinate system (u, v) whose coordinate curves are the lines of curvature. Let the curves $u = t$, $v = $ const be the lines of curvature passing along the principal vectors corresponding to zero principal curvature. Then by Rodrigues's theorem, $\vec{n}_u \equiv 0$, and hence, $(\vec{n}_u)_v \equiv 0$. Changing the order of differentiation, we obtain $(\vec{n}_v)_u \equiv 0$. From the last equality it follows that the field $\vec{n}_v$ that is orthogonal to the plane curve $u = t$, $v = $ const is a field of parallel vectors. Hence, $u = t$, $v = $ const is a line segment.

Theorem 2.7.3. *A regular surface* Φ *of class* C^k ($k \geq 2$) *with zero Gaussian curvature is a ruled surface, and hence is either a cone or a cylinder, or it is formed by the family of tangent lines to some space curve.*

If the surface has nonzero mean curvature, then the statement of Theorem 2.7.3 is proved, because in this case the second principal curvature differs from zero. In the general case, the proof of Theorem 2.7.3 requires a more delicate approach, which we omit.

2.7.2 Asymptotic Curves

Definition 2.7.2. A tangent vector $\vec{\lambda} \in T\Phi_P$ at a point P is called an *asymptotic direction* if $II(\vec{\lambda}) = 0$; i.e., the normal curvature of a surface Φ in this direction is zero.

Definition 2.7.3. An *asymptotic curve* is a smooth curve γ on a surface Φ whose tangent vector is an asymptotic direction at each point.

From Euler's formula one can see that asymptotic directions exist only at hyperbolic or parabolic points on a regular surface Φ. Moreover, at hyperbolic points there are two such directions, and at parabolic points, there is one.

Equations of asymptotic curves result directly from their definition. If $u = u(t)$, $v = v(t)$ are parametric equations of an asymptotic curve, then the functions $u(t)$ and $v(t)$ must satisfy the differential equation

$$L \, (du/dt)^2 + 2M(du/dt) \cdot (dv/dt) + N \, (dv/dt)^2 = 0. \qquad (2.83)$$

From Lemma 2.1.1 it immediately follows that in a neighborhood of a hyperbolic point there is a coordinate system in which the coordinate curves are the asymptotic curves.

From (2.83) it follows that coordinate curves are asymptotic curves if and only if $L = N = 0$. Indeed, if $u = t$, $v = $ const is an asymptotic curve, then $L \cdot 1 + 2M \cdot 0 + N \cdot 0 = 0$ or $L = 0$, analogously, if $u = $ const, $v = t$ is an asymptotic curve, then $N = 0$. Conversely, if $L = N = 0$, then (2.83) takes the form

$$M(du/dt) \cdot (dv/dt) = 0. \qquad (2.84)$$

If $M \neq 0$, then (2.84) has two solutions: $u = f(t)$, $v = $ const and $u = $ const, $v = f(t)$; i.e., the coordinate curves are the asymptotic curves. If $M = 0$, then a surface Φ in a neighborhood of the point under discussion is a plane, and each of its curves is an asymptotic curve.

We now study the geometrical characteristic of asymptotic curves.

Theorem 2.7.4. *If* γ *is an asymptotic curve on a regular surface* Φ *of class* C^k ($k \geq 3$), *then a tangent plane to* Φ *is an osculating plane of a curve* γ *at each point* $\gamma(t)$.

Proof. Consider an arbitrary point $\gamma(t)$ on the curve γ. Since the normal curvature of Φ in the direction of the tangent vector $\vec{\tau} = \dot{\gamma}(t)$ is zero, then by Meusnier's theorem, either the curvature of γ is zero at this point, or the angle between the principal normal $\vec{v}(t)$ to γ and the normal $\vec{n}(t)$ to Φ at the point $\gamma(t)$ is $\frac{\pi}{2}$. In the first case, any plane containing $\vec{\tau}(t)$ is an osculating plane, including $T\Phi_P$, and in the second case, the plane $T\Phi_P$ contains both $\vec{\tau}(t)$ and $\vec{v}(t)$, and hence it is again osculating.

Consider in detail a vector field $\vec{n}(t)$ along the asymptotic curve $\gamma(t)$. For those points where the torsion of $\gamma(t)$ exists, the vector $\vec{n}(t)$ equals $\pm\vec{\beta}(t)$, and hence $\kappa^2 = \left|\frac{d\vec{n}}{dt}\right|^2$; and for those points where the curvature of γ is zero and the torsion is not determined, we define it in addition, supposing that $\kappa^2 = \left|\frac{d\vec{n}}{dt}\right|^2$. Here t is the arc length parameter of a curve. $\square$

Keeping in mind this remark, we formulate the Beltrami–Enneper theorem.

Theorem 2.7.5. *The square of the torsion of an asymptotic curve on a regular surface of class C^k ($k \geq 3$) at each of its points is equal to the Gaussian curvature of the surface at this point, considered with the opposite sign.*

Proof. By definition,

$$\kappa^2 = |d\vec{n}/dt|^2 = III(\vec{\tau}),$$

where $\vec{\tau}$ is the tangent vector to the asymptotic curve. By (2.45) we have $III(\vec{\tau}) = -K \cdot I(\vec{\tau}) + 2H \cdot II(\vec{\tau})$, but by the condition of the theorem, $II(\vec{\tau}) = 0$, in view of the choice of parameter t: $I(\vec{\tau}) = 1$. Thus, we obtain $\kappa^2 = (\frac{d\vec{n}}{dt})^2 = III(\vec{\tau}) = -K$. The theorem is proved. $\square$

It is interesting to consider the particular case of the Beltrami–Enneper theorem in which an asymptotic curve γ is a straight line. Draw along $\gamma(t)$ on the surface Φ the unit vector field $\vec{\lambda}(t)$ orthogonal to $\gamma(t)$,

$$\vec{\lambda}(t) \in T\Phi_{\gamma(t)}, \quad \langle\vec{\lambda}(t), \vec{\tau}(t)\rangle = 0.$$

Here we again assume that t is an arc length parameter on γ. Write down the equalities

$$\left\langle\vec{\lambda}(t), \frac{d}{dt}\vec{\lambda}\right\rangle = 0, \tag{2.85a}$$

$$\langle\vec{\lambda}(t), \vec{\tau}(t)\rangle = 0, \tag{2.85b}$$

$$\vec{\lambda}(t) = \pm\vec{\tau}(t) \times \vec{n}(t), \tag{2.85c}$$

$$\vec{n}(t) = \pm\vec{\tau}(t) \times \vec{\lambda}(t), \tag{2.85d}$$

$$\frac{d\vec{\tau}}{dt} = 0. \tag{2.85e}$$

From (2.85c–e) we have

$$\frac{d}{dt}\vec{\lambda} = \pm\vec{\tau} \times \frac{d\vec{n}}{dt}, \qquad \frac{d\vec{n}}{dt} = \pm\vec{\tau} \times \frac{d}{dt}\vec{\lambda}, \tag{2.86}$$

and from this it follows that

$$\left\langle \frac{d\vec{n}}{dt}, \vec{\tau} \right\rangle = 0. \tag{2.87}$$

From (2.85e) and (2.86) we obtain

$$\left| \frac{d}{dt}\vec{\lambda} \right| = |\vec{\tau}| \cdot \left| \frac{d\vec{n}}{dt} \right| \cdot \sin\frac{\pi}{2} = \left| \frac{d\vec{n}}{dt} \right|.$$

Consequently,

$$\left(\frac{d}{dt}\vec{\lambda} \right)^2 = \left(\frac{d\vec{n}}{dt} \right)^2 = \kappa^2 = -K.$$

We have thus proved the following theorem.

Theorem 2.7.6. *Let a regular surface Φ of class C^k ($k \geq 3$) contain a straight line γ, and let $\vec{\lambda}(t)$ be the field of unit vectors along the straight line $\gamma(t)$ that are tangent to Φ and orthogonal to γ. Then the Gaussian curvature K is equal to $-|\frac{d}{dt}\vec{\lambda}|^2$, where t is the arc length of γ.*

2.7.3 Geodesics on a Surface

Here we give the definition and study the simplest properties of the most beautiful class of curves on a surface: the class of geodesics. We first define the notion of geodesic curvature of a curve on a surface Φ. Let $\gamma(t)$ ($a \leq t \leq b$) be a regular curve of class C^2 on a regular surface Φ of class C^2. Let t be the arc length of the curve $\gamma(t)$, counting from one of its points.

Denote, as usual, by $k(t)$ the curvature of γ at the point $\gamma(t)$, by $\vec{\tau}(t)$ and $\vec{\nu}(t)$ the tangent vector and principal normal vector at the point $\gamma(t)$, and by $\varphi(t)$ the angle between $\vec{\nu}(t)$ and $\vec{n}(t)$. The *geodesic curvature* $k_g(t)$ of the curve γ at the point $\gamma(t)$ is defined by the formula

$$k_g(t) = k(t) \cdot \sin\varphi(t). \tag{2.88}$$

In other words, the *geodesic curvature* $k_g(t)$ of γ at the point $\gamma(t)$ is the norm of the projection of the vector $k(t)\vec{\nu}(t)$ onto the tangent plane $T\Phi_{\gamma(t)}$ to Φ. Recall for comparison that the normal curvature of the curve γ (the normal curvature of Φ in the direction $\vec{\tau}$) is the projection of $k(t)\vec{\nu}(t)$ onto $\vec{n}$. Find the formula for calculation of k_g. From (2.88) and Theorem 1.6.1 it follows that

$$k_g(t) = k(t) \cdot \sin\varphi(t) = k(t) \cdot |\langle \vec{\nu} \times \vec{n}, \vec{\tau} \rangle| = (k(t)\vec{\nu} \cdot \vec{\tau} \cdot \vec{n}) = (\vec{r}'' \cdot \vec{r}' \cdot \vec{n}).$$

It is not difficult to derive that if t is an arbitrary parameter, then

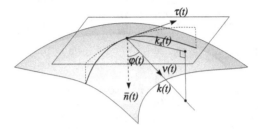

Figure 2.16. Geodesic curvature.

$$k_g(t) = \frac{(\vec{r}'' \cdot \vec{r}' \cdot \vec{n})}{|\vec{r}'|^3}. \tag{2.89}$$

We give now the definition of a geodesic.

Definition 2.7.4. A regular curve $\gamma(t)$ of class C^2 on a regular surface Φ of class C^2 is called a *geodesic* if its geodesic curvature at each point is zero.

From this definition it immediately follows that a normal $\vec{n}$ to a surface Φ along a geodesic $\gamma(t)$ coincides with the principal normal $\vec{v}(t)$ everywhere it is defined. Note that this property can be considered as the definition of a geodesic.

We deduce the equation of a geodesic. From (2.89) it follows that for a geodesic,

$$(\vec{r}'' \cdot \vec{r}' \cdot \vec{n}) = 0. \tag{2.90}$$

We calculate $\vec{r}'$ and $\vec{r}''$:

$$\vec{r}' = \vec{r}_u u' + \vec{r}_v v',$$
$$\vec{r}'' = \vec{r}_{uu}(u')^2 + 2\vec{r}_{uv}u'v' + \vec{r}_{vv}(v')^2 + \vec{r}_u u'' + \vec{r}_v v''. \tag{2.91}$$

Since the surface Φ is regular, then at any of its points the vectors $\vec{r}_u, \vec{r}_v, \vec{n}$ form a basis, and hence the vectors $\vec{r}_{uu}, \vec{r}_{uv}, \vec{r}_{vv}$ can be expressed as linear combinations of $\vec{r}_u, \vec{r}_v$, and $\vec{n}$. Let

$$\begin{cases} \vec{r}_{uu} = \Gamma_{11}^1 \vec{r}_u + \Gamma_{11}^2 \vec{r}_v + \alpha_{11}\vec{n}, \\ \vec{r}_{uv} = \Gamma_{12}^1 \vec{r}_u + \Gamma_{12}^2 \vec{r}_v + \alpha_{12}\vec{n}, \\ \vec{r}_{vv} = \Gamma_{22}^1 \vec{r}_u + \Gamma_{22}^2 \vec{r}_v + \alpha_{22}\vec{n}. \end{cases}$$

The geometrical sense of the coefficients in the above expressions will become clear later on, in Section 2.8 (and in Chapter 3), and there the expression of these coefficients will be given through the coefficients of the first and the second fundamental forms. Substituting the last equations into (2.91), and then into (2.90), we obtain

$$\vec{r}'' = (u'' + A)\vec{r}_u + (v'' + B)\vec{r}_v + c\vec{n},$$

where

$$A = \Gamma^1_{11}(u')^2 + 2\Gamma^1_{12}u'v' + \Gamma^1_{22}(v')^2,$$
$$B = \Gamma^2_{11}(u')^2 + 2\Gamma^2_{12}u'v' + \Gamma^2_{22}(v')^2. \tag{2.92}$$

From (2.92) and (2.90) it follows that

$$u''v' - u'v'' + Av' - Bu' = 0. \tag{2.93}$$

We prove the following important local theorem.

Theorem 2.7.7. *Through every point on a regular surface Φ of class C^2 in any direction one can pass one and only one geodesic.*

Proof. Let $P_0(u_0, v_0)$ be any point on Φ, and let $\vec{\lambda}_0 = \lambda^1_0 \vec{r}_u + \lambda^2_0 \vec{r}_v$ be any nonzero vector at P_0; hence $\vec{\lambda}_0 \in T\Phi_{P_0}$. Let for definiteness, $\lambda^1_0 \neq 0$. If the equation of a geodesic is given explicitly as $u = u$, $v = v(u)$, then (2.93) takes the form

$$-v'' + \bar{A}v' - \bar{B} = 0,$$

where

$$\bar{A} = \Gamma^1_{11} + 2\Gamma^1_{12}v' + \Gamma^1_{22}(v')^2, \quad \bar{B} = \Gamma^2_{11} + 2\Gamma^2_{12}v' + \Gamma^2_{22}(v')^2. \tag{2.94}$$

The initial conditions for the function $v(u)$ will be the following:

$$v(u) = v_0, \quad \frac{dv}{du} = v'(u_0) = \frac{\lambda^2_0}{\lambda^1_0}. \tag{2.95}$$

From (2.85d) and (2.95), using the standard existence and uniqueness theorem for a solution of ordinary differential equations, the statement of the theorem follows.   □

We consider some examples.

Example 2.7.1 (Geodesics on a sphere). Directly from the definition it follows that the *great circles* on a sphere are actually the geodesics, since a normal to a sphere and the principal normal to a great circle at all points are parallel, and from Theorem 2.7.7 it follows that there are no other geodesics, since we can pass a great circle through each point on a sphere in any direction.

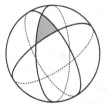

Figure 2.17. Geodesics on a sphere and a cylinder.

Example 2.7.2 (Geodesics on a cylinder). The equation of a *right circular cylinder C* of radius a can be written in parametric form as

$$x = a \cos u, \quad y = a \sin u, \quad z = v,$$

for $-\infty < u < \infty, -\infty < u < \infty$. Prove that any *helix* lying on C is a geodesic.
Let

$$\vec{r} = \vec{r}(t) = a \cos(t - t_0)\vec{i} + a \sin(t - t_0)\vec{j} + [b(t - t_0) + c]\vec{k}$$

be the equation of a helix (the equation of a *parallel* for $b = 0$). Then

$$\vec{r}'(t) = -a \sin(t - t_0)\vec{i} + a \cos(t - t_0)\vec{j} + b\vec{k},$$
$$\vec{r}''(t) = -a \cos(t - t_0)\vec{i} - a \sin(t - t_0)\vec{j},$$

and

$$\vec{r}_u = -a \sin u \vec{i} + a \cos u \vec{j}, \quad \vec{r}_v = \vec{k},$$

but

$$\vec{r}_u \times \vec{r}_v = a \cos u \vec{i} + a \sin u \vec{j},$$

and we see that the vectors $\vec{r}''(t)$ and $\vec{n}(u, v)$ are parallel at points of a helix. Consequently, $(\vec{r}'' \cdot \vec{r}' \cdot \vec{n}) = 0$, and hence a helix for every b, c, and t_0 is a geodesic on the cylinder C. Note that the *rulings* of the cylinder C (i.e., the straight lines $x = x_0, y = y_0, z = t - t_0$) are also the geodesics.

We prove that there are no other geodesics on the cylinder C. In view of Theorem 2.7.7, it is sufficient to show that through any point $P_0(u_0, v_0)$ on the cylinder C and in any direction $\vec{\lambda} = \lambda^1 \vec{r}_u + \lambda^2 \vec{r}_v$ one can pass a helix if $\vec{\lambda} \neq 0$, and if $\vec{\lambda} = \vec{r}_v$, then a ruling. Indeed, let

$$\vec{r}(t) = a \cos(t - t_0)\vec{i} + a \sin(t - t_0)\vec{j} + [b(t - t_0) + c]\vec{k},$$

where

$$t_0 = -u_0, \quad b = \frac{\lambda^2}{\lambda^1}u_0, \quad c = v_0 - \frac{\lambda^2}{\lambda^1}v_0.$$

Then

$$\vec{r}(0) = a \cos(u_0)\vec{i} + \lambda a \cos(u_0)\vec{j} + \lambda^1\vec{k} = \lambda^1\vec{r}_u + \lambda^2\vec{r}_v,$$

which completes the proof.

Example 2.7.3 (Geodesics on a surface of revolution). Let Φ be the regular surface of class C^2 obtained by rotation of a curve γ around the axis OZ (see Section 2.6.1). Directly from the main property of geodesics it follows that all meridians on Φ are geodesics. Usually, one cannot find other geodesics in the form of explicit equations. But it is possible to point out a property of geodesics that allows us to give a qualitative view of their behavior on a surface of revolution.

Let $\vec{r} = \vec{r}(t)$ be the arc length parameterization of a geodesic $\sigma(t)$ on Φ. Denote by $\rho(t)$ the distance from a point $\sigma(t)$ to the axis of rotation, and by $\alpha(t)$ the angle between $\sigma(t)$ and a parallel at the point of their intersection, $\sigma(t)$. Then the equality

$$\rho(t) \cdot \cos \alpha(t) = c \qquad (2.96)$$

(*Clairaut's theorem*) is satisfied. First, we shall prove the equality

$$\rho(t) \cdot \cos \alpha(t) = (\vec{r} \cdot \vec{r}' \cdot \vec{k}), \qquad (2.97)$$

where $\vec{k}$ is a unit basis vector of the axis of rotation OZ. Denote by $\vec{\mu}(t)$ a unit vector tangent to a parallel of the surface Φ at the point $\sigma(t)$. Then

$$\vec{k} \times \vec{r}(t) = \rho(t)\vec{\mu}(t), \quad \langle \vec{\mu}(t), \vec{r}'(t) \rangle = \cos \alpha(t). \qquad (2.98)$$

From equalities (2.98) follows (2.97). Differentiating the right-hand side of (2.97) with respect to t, we obtain

$$(\vec{r} \cdot \vec{r}' \cdot \vec{k})' = (\vec{r} \cdot \vec{r}'' \cdot \vec{k}). \qquad (2.99)$$

Since t is the arc length of $\sigma(t)$, then $\vec{r}''(t)$ is parallel to the principal normal $\vec{v}(t)$ of the curve $\sigma(t)$, and since $\sigma(t)$ is a geodesic, then $\vec{v}(t)$ is parallel to the normal $\vec{n}(t)$ of Φ at the point $\sigma(t)$. Consequently, $\vec{r}''(t)$ is parallel to $\vec{n}(t)$. But on the other hand, since Φ is a surface of revolution, then $\vec{n}(t)$ lies in the plane spanned by $\vec{k}$ and $\vec{r}(t)$. Consequently, the vectors $\vec{k}, \vec{r}(t)$, and $\vec{r}''(t)$ are coplanar, and hence

$$(\vec{r}(t) \cdot \vec{r}''(t) \cdot \vec{k}) = 0.$$

From this equality and equality (2.97) follows our statement, the equality (2.96). The equality (2.96) gives us the possibility to describe the "qualitative" behavior of a geodesic. Let $c \neq 0$ (if $c = 0$, then we obtain that σ is a meridian). When we move along a geodesic $\sigma(t)$, passing from "wider" part of Φ into its "thinner" part, then the functions $\rho(t)$ and $\alpha(t)$ decrease; i.e., the angle between the geodesic $\sigma(t)$ and the parallels on Φ becomes less and less. If for some t the value of $\rho(t)$ becomes equal to c, then at this point $\sigma(t)$ touches a parallel, and after this returns into the "wider" part of Φ. While passing from the "thinner" part on the surface Φ into the "wider" part, the functions $\rho(t)$ and $\alpha(t)$ increase, and consequently, the angle between $\sigma(t)$ and the meridians becomes less and less. If the "wide" part becomes limitlessly "wide," then the direction of a geodesic limitlessly tends to the direction of a meridian; see Figure 2.18.

Finally, if a tangent line to the curve γ at some point is parallel to the axis of rotation, then the parallel on Φ corresponding to this point becomes a closed geodesic, because in this case the normal to the surface Φ is at the same time the principal normal to this parallel.

2.7.4 Problems

The following two topological lemmas are necessary for us to solve a sequence of problems.

Let D be some region homeomorphic to a disk on a regular surface Φ with a regular boundary $\partial D = \gamma$, and let $\vec{e}(P)$ be a continuous vector field in D.

Figure 2.18. Geodesics on a surface of revolution.

Lemma 2.7.1. If for each point Q on the curve γ the vector $\vec{e}(Q)$ forms a nonzero angle with a tangent vector $\dot{\gamma}$ at a point $\gamma(t)$, then there is a point $Q_0 \in \operatorname{int} D$ at which $\vec{e}(Q_0) = 0$ holds.

Lemma 2.7.2. Let Φ be a regular surface, homeomorphic to a sphere. Then for any continuous field $\vec{e}(P)$ on Φ (i.e., tangent) there is a point Q with the property $\vec{e}(Q) = 0$.

Remark 2.7.1. Lemma 2.7.2 is a particular case of the *Poincaré–Brouwer theorem: If Φ is a compact surface and X is a vector field on Φ with only a finite number of zeros, then the total index $I(X) = \sum_i i_p(X)$ is $2\pi \chi(\Phi)$, where $\chi(\Phi)$ is the Euler characteristic of Φ.* Here $i_p(X)$ is the angular variation of X along "small" circle on Φ with center at P. The Euler characteristic of the sphere is 2.

Problem 2.7.1. Let Φ be a regular surface of class C^2. Prove that

(1) if a line of curvature is a geodesic, then it is a plane curve, or more exactly, the torsion of this curve is zero at every point where it (the torsion) is defined;
(2) if an asymptotic curve is at the same time a geodesic, then it is a straight line;
(3) if a line of curvature is at the same time an asymptotic curve, then it is a plane curve.

Hint. Use Rodrigues's theorem 2.4.5, the properties of asymptotic curves and geodesics, and the Frenet formulas from Chapter 1.

Problem 2.7.2. On any regular surface Φ of class C^2 homeomorphic to a sphere there is at least one umbilic.

Solution. Assume the opposite. This means that we assume the inequality

$$k_1(P) < k_2(P), \tag{2.100}$$

where as usual, $k_1(P)$, $k_2(P)$ are the principal curvatures of Φ at the point P. Let $P_0 \in \Phi$ and let a unit vector $\vec{e}$ be parallel to a principal direction that corresponds to the principal curvature $k_1(P)$. Build a vector field of unit vectors on the whole surface Φ by the following method: join the point P_0 with an arbitrary point P by some continuous curve

$$\sigma(t) \quad (0 \le t \le 1, \quad \sigma(0) = P_0, \quad \sigma(1) = P),$$

and let $\vec{e}(t)$ be a unit continuous vector field along $\sigma(t)$ with the condition $\vec{e}(0) = \vec{e}(P)$, and $\vec{e}(t)$ for each t is parallel to a principal direction. In view of the inequality (2.100), a vector field $\vec{e}(t)$ is uniquely determined by these conditions. Now assume $\vec{e}(P) = \vec{e}(1)$. We prove that the definition of the vector $\vec{e}(P)$ does not depend on the choice of curve $\sigma(t)$ joining P_0 and P. Let

$$\sigma_1(t) \quad (0 \le t \le 1, \quad \sigma_1(0) = P_0, \quad \sigma_1(1) = P)$$

be some other continuous curve relating P_0 with the point P. Repeating the previous construction for the curve σ_1, we obtain a vector $\vec{e}_{\sigma_1}(P)$, which is either equal to $\vec{e}(P)$ or to $\vec{e}(P) = -\vec{e}_{\sigma_1}(P)$. But since Φ is homeomorphic to a sphere, there is on it a continuous deformation $\sigma_u(t)$ $(0 \le u \le 1)$ of the path σ_1 onto the path σ, $\sigma(0, t) = \sigma_0, \sigma(1, t) = \sigma_1$, for which the vector $\vec{e}_{\sigma_u}(P)$ continuously depends on u. Thus $\vec{e}_{\sigma(1)}(P) = \vec{e}_{\sigma(0)}(P) = \vec{e}(P)$. So, we have built a continuous unit vector field $\vec{e}(P)$ on the whole surface Φ, which contradicts Lemma 2.7.2. $\qquad\square$

Problem 2.7.3. If on a regular surface Φ of class C^2 there is a closed line of curvature bounding on Φ a region D homeomorphic to a circle, then there is an umbilic in D.

Hint. Repeat with obvious modifications the proof of Problem 2.7.2 and apply Lemma 2.7.1.

Problem 2.7.4. On a regular surface Φ of class C^2 with negative Gaussian curvature there are no closed asymptotic curves.

Hint. The solution is analogous to the solution of Problem 2.7.3.

Problem 2.7.5.* If on a saddle surface homeomorphic to a plane there is a closed asymptotic curve, then the region bounded by this curve, is a region on the plane.

This assertion was formulated by A.V. Pogorelov, but it has not yet been proved.

Problem 2.7.6. Let Φ be a regular surface of class C^2 and $\gamma(t)$ $(a \le t \le b)$ a line of curvature on Φ. Then if the Gaussian curvature of Φ is either negative or positive,

$$\left| \int_{t_0}^{t_1} \kappa(t)\, dt \right| < \pi$$

holds for all $t_0, t_1 \in (a, b)$, where κ is the torsion and t is the arc length of the curve. Note that the case $a = -\infty$ and $b = \infty$ is not excluded.

Solution. From the restrictions given for the Gaussian curvature of Φ it follows that the principal curvatures $k_1(P)$ and $k_2(P)$ at each point P differ from zero, and consequently, by Meusnier's theorem the curvature $k(t)$ of the curve $\gamma(t)$ is everywhere positive:

$$k(t) > 0 \quad (a < t < b). \tag{2.101}$$

Let $\vec{\tau}(t)$, $\vec{v}(t)$, $\vec{\beta}(t)$, and $\vec{n}(t)$ be four vector fields along γ, where $\vec{\tau}(t)$ is a tangent vector field, $\vec{v}(t)$ is the field of principal normal vectors, $\vec{\beta}(t)$ binormals, and $\vec{n}(t)$ is a field of normals to Φ along $\gamma(t)$ such that $k_1(t) > 0$. Compose $\vec{n}(t)$ as a linear combination of $\vec{v}(t)$ and $\vec{\beta}(t)$:

$$\vec{n}(t) = \cos\alpha(t)\vec{v}(t) + \sin\alpha(t)\vec{\beta}(t). \tag{2.102}$$

Then, on the one hand, by Rodrigues's theorem,

$$\vec{n}'(t) = -k_1(t)\vec{\tau}(t), \tag{2.103}$$

and on the other hand, from (2.102) and the Frenet formulas of Chapter 1, we obtain

$$\begin{aligned} \vec{n}'(t) = &- \sin\alpha(t) \cdot \alpha'(t)\vec{v}(t) + \cos\alpha(t)(-k\vec{\tau}(t) - \kappa\vec{\beta}(t)) \\ &+ \cos\alpha(t) \cdot \alpha'(t)\vec{\beta}(t) + \sin\alpha(t)\kappa\vec{v}(t) = -k\cos\alpha(t)\vec{\tau}(t) \\ &- \sin\alpha(t)(\alpha'(t) - \kappa)\vec{v}(t) + \cos\alpha(t)(\alpha'(t) - \kappa)\vec{\beta}(t). \end{aligned} \tag{2.104}$$

Comparing (2.103) with (2.104), we have the system

$$\begin{cases} k_1(t) = k \cdot \cos\alpha(t), \\ \sin\alpha(t)(\alpha'(t) - \kappa) = 0, \\ \cos\alpha(t)(\alpha'(t) - \kappa) = 0. \end{cases} \tag{2.105}$$

From (2.105) it follows that

$$\alpha'(t) = \kappa(t). \tag{2.106}$$

Let t_0 be an arbitrary real number from the interval (a, b), and $\alpha(t_0) = \alpha_0$. Then from (2.106) follows

$$\alpha(t) = \int_{t_0}^{t} \kappa(t)\,dt + \alpha_0. \tag{2.107}$$

Finally, from (2.101) and from the conditions $k_1(t) > 0$ and (2.105) we obtain that

$$\cos\alpha(t) = k_1(t)/k(t) > 0.$$

Hence $\alpha(t)$ satisfies the inequalities

$$-\frac{\pi}{2} < \alpha(t) < \frac{\pi}{2}. \tag{2.108}$$

In particular, $-\frac{\pi}{2} < \alpha_0 < \frac{\pi}{2}$. From (2.107) and (2.108) the statement of the problem follows. $\qquad\square$

Let γ be a regular curve in $\mathbb{R}^3$ of class C^3, and $\gamma(u)$ its natural parameterization, $c < u < d$ ($c < u \le d$). In the case $c < u \le d$, suppose that γ is closed, and in the first case, the equalities $c = \infty$ and $d = \infty$ are not excluded. In each normal plane to γ at a point $\gamma(t)$ take the circle of radius a with center at the point $\gamma(t)$. The obtained one-parameter family of circles forms, generally speaking, a regular surface of class C^2, which we denote by $\Phi(a)$ and call a *generalized cylinder* or if Φ is closed, a *generalized torus*. Denote by $k(u)$ and $\kappa(u)$ the curvature and the torsion of γ at the point $\gamma(t)$.

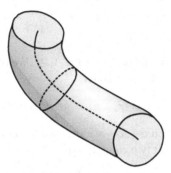

Figure 2.19. Generalized cylinder.

Problem 2.7.7. Prove that a surface $\Phi(a)$ is regular if and only if $k(u) < \frac{1}{a}$ holds. Prove that there are no umbilics on a regular surface $\Phi(a)$. Prove that the largest principal curvature of a surface $\Phi(a)$ is constant and equal to $\frac{1}{a}$. Find the equations for the lines of curvature on $\Phi(a)$.

Hint. The problem can be solved by direct computation if the parameterization of $\Phi(a)$ is given in the following form:

$$\vec{r} = \vec{r}(u, v) = \vec{\rho}(u) + a(\cos\alpha(u, v)\vec{v} + \sin\alpha(u, v)\vec{\beta}(t)),$$

where $\vec{\rho}(u)$ is the parameterization of γ,

$$\alpha(u, v) = v + \int_{u_0}^{u} \kappa(u)\,du,$$

$c < u < d$ ($c < u \le d$), and

$$\int_{u_0}^{u} \kappa(u)\,du < v \le 2\pi + \int_{u_0}^{u} \kappa(u)\,du.$$

2.8 The Main Equations of Surface Theory

2.8.1 Derivational formulas

At each point P on a regular parameterized surface $\Phi: \vec{r} = \vec{r}(u, v)$ of class C^k ($k \ge 2$), the three vectors $\vec{r}_u$, $\vec{r}_v$, and $\vec{n}$ are defined. These three vectors, in view of the regularity of Φ, are linearly independent and consequently form a basis. Thus, any other vector can be presented as a linear combination of $\vec{r}_u$, $\vec{r}_v$, and $\vec{n}$. Find the coefficients of this decomposition by the basis of the vectors $\vec{r}_{uu}$, $\vec{r}_{uv}$, $\vec{r}_{vv}$ and $\vec{n}_u$, $\vec{n}_v$. Let

$$
\begin{aligned}
\vec{r}_{uu} &= \Gamma_{11}^1 \vec{r}_u + \Gamma_{11}^2 \vec{r}_v + A_{11}\vec{n}, \\
\vec{r}_{uv} &= \Gamma_{12}^1 \vec{r}_u + \Gamma_{12}^2 \vec{r}_v + A_{12}\vec{n}, \\
\vec{r}_{vv} &= \Gamma_{22}^1 \vec{r}_u + \Gamma_{22}^2 \vec{r}_v + A_{22}\vec{n}, \\
\vec{n}_u &= \alpha_{11}\vec{r}_u + \alpha_{12}\vec{r}_v, \\
\vec{n}_v &= \alpha_{21}\vec{r}_u + \alpha_{22}\vec{r}_v.
\end{aligned}
\tag{2.109}
$$

It is easiest to find the coefficients A_{ij} ($i, j = 1, 2$). We obtain the scalar product of the first three equations of (2.109) on $\vec{n}$:

$$
A_{11} = \langle \vec{r}_{uu}, \vec{n} \rangle = L, \quad A_{12} = \langle \vec{r}_{uv}, \vec{n} \rangle = M, \quad A_{22} = \langle \vec{r}_{vv}, \vec{n} \rangle = N. \tag{2.110}
$$

For coefficients α_{11}, α_{12} and α_{21}, α_{22} we obtain a system of equations by taking the scalar product of the last two equations of (2.109) with $\vec{r}_u$ and $\vec{r}_v$:

$$
\begin{cases} \alpha_{11}E + \alpha_{12}F = -L \\ \alpha_{11}F + \alpha_{12}G = -M, \end{cases} \tag{2.111}
$$

$$
\begin{cases} \alpha_{21}E + \alpha_{22}F = -M \\ \alpha_{21}F + \alpha_{22}G = -N. \end{cases} \tag{2.112}
$$

From (2.111) and (2.112) we have

$$
\begin{aligned}
\alpha_{11} &= \frac{-LG + MF}{EG - F^2}, \quad &\alpha_{12} &= \frac{LF - ME}{EG - F^2}, \\
\alpha_{21} &= \frac{NF - MG}{EG - F^2}, \quad &\alpha_{22} &= \frac{-NE + MF}{EG - F^2}.
\end{aligned}
\tag{2.113}
$$

It is interesting to note that for a coordinate system whose coordinate curves coincide with the lines of curvature, equations (2.113) can be rewritten in the form

$$
\alpha_{11} = -k_1 \quad \alpha_{12} = \alpha_{21} = 0, \quad \alpha_{22} = -k_2,
$$

where k_1, and k_2 are the principal curvatures of Φ at a given point. We shall obtain a system of equations for the coefficients Γ_{jk}^i, applying the scalar multiplication of the first three equations of (2.109) on $\vec{r}_u$ and $\vec{r}_v$:

$$
\begin{cases} \Gamma_{11}^1 E + \Gamma_{11}^2 F = \langle \vec{r}_{uu}, \vec{r}_u \rangle = \frac{1}{2}E_u \\ \Gamma_{11}^1 F + \Gamma_{11}^2 G = \langle \vec{r}_{uu}, \vec{r}_v \rangle = F_u - \frac{1}{2}E_v, \end{cases} \tag{2.114}
$$

$$
\begin{cases} \Gamma_{12}^1 E + \Gamma_{12}^2 F = \langle \vec{r}_{uv}, \vec{r}_u \rangle = \frac{1}{2}E_v \\ \Gamma_{12}^1 F + \Gamma_{12}^2 G = \langle \vec{r}_{uv}, \vec{r}_v \rangle = \frac{1}{2}G_u, \end{cases} \tag{2.115}
$$

$$
\begin{cases} \Gamma_{22}^1 E + \Gamma_{22}^2 F = \langle \vec{r}_{vv}, \vec{r}_u \rangle = F_v - \frac{1}{2}G_u \\ \Gamma_{22}^1 F + \Gamma_{22}^2 G = \langle \vec{r}_{vv}, \vec{r}_v \rangle = \frac{1}{2}G_v. \end{cases} \tag{2.116}
$$

By the way, we have used the identities

$$\langle \vec{r}_{uu}, \vec{r}_v \rangle = -\frac{\partial}{\partial u}\langle \vec{r}_u, \vec{r}_v \rangle - \langle \vec{r}_u \vec{r}_{uv} \rangle, \quad \langle \vec{r}_{vv}, \vec{r}_u \rangle = \frac{\partial}{\partial v}\langle \vec{r}_v, \vec{r}_u \rangle - \langle \vec{r}_v \vec{r}_{uv} \rangle.$$

Solving the systems (2.114)–(2.116) we obtain

$$
\begin{aligned}
\Gamma_{11}^1 &= \frac{\frac{1}{2}E_u G + \frac{1}{2}E_v F - F F_u}{EG - F^2}, & \Gamma_{11}^2 &= \frac{F_u E - \frac{1}{2}E_v E - \frac{1}{2}E_u F}{EG - F^2}, \\
\Gamma_{12}^1 &= \frac{\frac{1}{2}E_v G - \frac{1}{2}G_u F}{EG - F^2}, & \Gamma_{12}^2 &= \frac{\frac{1}{2}G_u E - \frac{1}{2}E_v F}{EG - F^2}, \\
\Gamma_{22}^1 &= \frac{-\frac{1}{2}G_u G - \frac{1}{2}G_v F + G F_v}{EG - F^2}, & \Gamma_{22}^2 &= \frac{\frac{1}{2}G_v E + \frac{1}{2}G_u F - F F_v}{EG - F^2}.
\end{aligned}
\tag{2.117}
$$

If the coordinate system (u, v) is orthogonal, i.e., $F(u, v) = 0$, then the formulas for Γ_{jk}^i $(i, j, k = 1, 2)$ are essentially simplified:

$$
\begin{aligned}
\Gamma_{11}^1 &= \frac{E_u}{2E}, & \Gamma_{11}^2 &= -\frac{E_v}{2G}, & \Gamma_{12}^1 &= \frac{E_v}{2E}, \\
\Gamma_{12}^2 &= \frac{G_u}{2G}, & \Gamma_{22}^1 &= -\frac{G_u}{2E}, & \Gamma_{22}^2 &= \frac{G_v}{2G}.
\end{aligned}
\tag{2.118}
$$

The geometrical sense of the coefficients Γ_{jk}^i $(i, j, k = 1, 2)$ will be cleared up below, in Chapter 3. Now note only that they are expressed through the coefficients of the first fundamental form and their first derivatives only. The coefficients themselves are called the *Christoffel symbols of the second kind*. Finally, note that the *derivational formulas* (2.109) can be considered as a direct generalization of the Frenet formulas for the space curves (see Chapter 1).

2.8.2 Gauss–Peterson–Codazzi Formulas

If $E(\lambda^1)^2 + 2F(\lambda^1)(\lambda^2) + G(\lambda^2)^2$ and $L(\lambda^1)^2 + 2M(\lambda^1)(\lambda^2) + N(\lambda^2)^2$ are the first and the second fundamental forms of a surface, then they cannot be taken arbitrarily. There are definite relations between the coefficients of these fundamental forms. These relations can be found based on the independence of the derivatives from the order of derivation. If a regular surface Φ belongs to class C^k $(k \geq 3)$ and $\vec{r} = \vec{r}(u, v)$ is its vector equation, then the following equations hold:

$$(\vec{r}_{uu})_v = (\vec{r}_{uv})_u, \quad (\vec{r}_{vv})_u = (\vec{r}_{uv})_v, \quad (\vec{n}_u)_v = (\vec{n}_v)_u.$$

If we substitute in these formulas the expressions for $\vec{r}_{uu}, \vec{r}_{uv}, \vec{r}_{vv}, \vec{n}_u, \vec{n}_v$ from (2.109), then the three vector equalities

$$
\begin{aligned}
(\Gamma_{11}^1 \vec{r}_u + \Gamma_{11}^2 \vec{r}_v + L\vec{n})_v &= (\Gamma_{12}^1 \vec{r}_u + \Gamma_{12}^2 \vec{r}_v + M\vec{n})_u, \\
(\Gamma_{22}^1 \vec{r}_u + \Gamma_{22}^2 \vec{r}_v + N\vec{n})_u &= (\Gamma_{12}^1 \vec{r}_u + \Gamma_{12}^2 \vec{r}_v + M\vec{n})_v, \\
(\alpha_{11}\vec{r}_u + \alpha_{12}\vec{r}_v)_v &= (\alpha_{21}\vec{r}_u + \alpha_{22}\vec{r}_v)_u.
\end{aligned}
\tag{2.119}
$$

are obtained. Differentiating them, we obtain

$$\frac{\partial \Gamma_{11}^1}{\partial v}\vec{r}_u + \Gamma_{11}^1 \vec{r}_{uv} + \frac{\partial \Gamma_{11}^2}{\partial v}\vec{r}_v + \Gamma_{11}^2 \vec{r}_{vv} + L_v \vec{n} + L\vec{n}_v$$

$$= \frac{\partial \Gamma_{12}^1}{\partial u}\vec{r}_u + \Gamma_{12}^1 \vec{r}_{uu} + \frac{\partial \Gamma_{12}^2}{\partial u}\vec{r}_v + \Gamma_{12}^2 \vec{r}_{uv} + M_u \vec{n} + M\vec{n}_u,$$

$$\frac{\partial \Gamma_{22}^1}{\partial u}\vec{r}_u + \Gamma_{22}^1 \vec{r}_{uu} + \frac{\partial \Gamma_{22}^2}{\partial u}\vec{r}_v + \Gamma_{22}^2 \vec{r}_{uv} + N_u \vec{n} + N\vec{n}_u$$

$$= \frac{\partial \Gamma_{12}^1}{\partial v}\vec{r}_u + \Gamma_{12}^1 \vec{r}_{uv} + \frac{\partial \Gamma_{12}^2}{\partial v}\vec{r}_v + \Gamma_{12}^2 \vec{r}_{vv} + M_v \vec{n} + M\vec{n}_v,$$

$$\frac{\partial \alpha_{11}}{\partial v}\vec{r}_u + \alpha_{11}\vec{r}_{uv} + \frac{\partial \alpha_{12}}{\partial v}\vec{r}_v + \alpha_{12}\vec{r}_{vv}$$

$$= \frac{\partial \alpha_{21}}{\partial u}\vec{r}_u + \alpha_{21}\vec{r}_{uu} + \frac{\partial \alpha_{22}}{\partial u}\vec{r}_v + \alpha_{22}\vec{r}_{uv}. \quad (2.120)$$

Substituting above the expressions for the vectors $\vec{r}_{uu}$, $\vec{r}_{uv}$, $\vec{r}_{vv}$ and $\vec{n}_u$, $\vec{n}_v$ from (2.109) and collecting together the coefficients at $\vec{r}_u$, $\vec{r}_u$, $\vec{n}$, we obtain equalities of the following form:

$$\begin{cases} \tilde{A}_{11}\vec{r}_u + \tilde{A}_{12}\vec{r}_v + B_1\vec{n} = 0, \\ \tilde{A}_{21}\vec{r}_u + \tilde{A}_{22}\vec{r}_v + B_2\vec{n} = 0, \\ \tilde{A}_{31}\vec{r}_u + \tilde{A}_{32}\vec{r}_v + B_3\vec{n} = 0. \end{cases} \quad (2.121)$$

From this, in view of the linear independence of the vectors $\vec{r}_u$, $\vec{r}_v$, $\vec{n}$, it follows that all coefficients in (2.121) are zero. Altogether, we have obtained nine scalar equalities. But it turns out that only three of them are independent, and the others become identities when these three are satisfied. We find an expression for $\tilde{A}_{12}$:

$$\tilde{A}_{12} = \frac{\partial \Gamma_{11}^2}{\partial v} - \frac{\partial \Gamma_{12}^2}{\partial u} + \Gamma_{11}^1 \Gamma_{12}^2 + \Gamma_{11}^2 \Gamma_{22}^2 - \Gamma_{12}^1 \Gamma_{11}^2 - \Gamma_{12}^2 \Gamma_{12}^2 + L\alpha_{22} - M\alpha_{12} = 0$$

or

$$M\alpha_{12} - L\alpha_{22} = \frac{\partial \Gamma_{11}^2}{\partial v} - \frac{\partial \Gamma_{12}^2}{\partial u} + \Gamma_{11}^1 \Gamma_{12}^2 + \Gamma_{11}^2 \Gamma_{22}^2 - \Gamma_{12}^1 \Gamma_{11}^2 - \Gamma_{12}^2 \Gamma_{12}^2$$

$$= T. \quad (2.122)$$

Substituting in (2.122) the expressions for α_{12} and α_{22}, we obtain

$$M\frac{LF - ME}{EG - F^2} - L\frac{-NE + MF}{EG - F^2} = T, \quad \text{or} \quad E\frac{LN - M^2}{EG - F^2} = T.$$

Finally, we obtain that the Gaussian curvature K of Φ is

$$K = \frac{1}{E}\left(\frac{\partial \Gamma_{11}^2}{\partial v} - \frac{\partial \Gamma_{12}^2}{\partial u} + \Gamma_{11}^1 \Gamma_{12}^2 + \Gamma_{11}^2 \Gamma_{22}^2 - \Gamma_{12}^1 \Gamma_{11}^2 - \Gamma_{12}^2 \Gamma_{12}^2\right). \quad (2.123)$$

The equality (2.123) was first obtained by Gauss and is called *Gauss's "Theorema Egregium"* which means *remarkable*. From this equality it is seen that the Gaussian curvature of a surface can be calculated when only the coefficients of the first fundamental form are given.[8] Hence Gaussian curvature is an object of the intrinsic geometry of a surface.

I think that actually, differential geometry has been distinguished as an independent subject only after the work of Gauss, and that Gauss's *"theorema egregium"* has played the main part in this process. As we shall see below, in Chapter 3, Gaussian curvature determines the most essential properties of a surface from the point of view of its intrinsic geometry. This theorem has also served as a starting point for the development of *n-dimensional Riemannian geometry*. From the remaining eight equations only two are linearly independent; the others are consequences of these. The last two equations were obtained by K. Peterson and later by D. Codazzi and G. Mainardi. They can be written in the following simple and symmetric form (by A.V. Pogorelov):

$$2(EG - F^2)(L_v - M_u)$$

$$- (EN - 2FM + GL)(E_v - F_u) + \begin{vmatrix} E & E_u & L \\ F & F_u & M \\ G & G_u & N \end{vmatrix} = 0, \quad (2.124)$$

$$2(EG - F^2)(M_v - N_u)$$

$$- (EN - 2FM + GL)(F_v - G_u) + \begin{vmatrix} E & E_v & L \\ F & F_v & M \\ G & G_v & N \end{vmatrix} = 0. \quad (2.125)$$

The formulas (2.124) and (2.125) are known as the *Peterson–Codazzi formulas*. Finally, *Gauss's formula* can be written in the form

$$K = \frac{-1}{2\sqrt{EG - F^2}}\left[\left(\frac{E_v - F_u}{\sqrt{EG - F^2}}\right)_v - \left(\frac{F_v - G_u}{\sqrt{EG - F^2}}\right)_u\right]$$

$$- \frac{1}{4(EG - F^2)^2}\begin{vmatrix} E & E_u & E_v \\ F & F_u & F_v \\ G & G_u & G_v \end{vmatrix}. \quad (2.126)$$

The Peterson–Codazzi formulas take a quite simple form if the coordinate system (u, v) is such that the *coordinate curves of this system are the lines of curvature*. In such a coordinate system, as we know, $F = M = 0$, and then (2.124) and (2.125) can be simplified to

---

[8] In other words, the Gaussian curvature of a surface is preserved by isometries.

$$L_v = \frac{E_v}{2}\left(\frac{L}{E} + \frac{N}{G}\right), \quad N_u = \frac{G_u}{2}\left(\frac{L}{E} + \frac{N}{G}\right). \tag{2.127}$$

The principal curvatures $k_1(u, v)$ and $k_2(u, v)$ in this coordinate system are

$$k_1(u, v) = \frac{L}{E}, \quad k_2(u, v) = \frac{N}{G}. \tag{2.128}$$

Thus (2.127) take the form

$$L_v = (k_1)_v E + k_1 E_v, \quad N_u = (k_2)_u G + k_2 G_u. \tag{2.129}$$

Substituting L_v and N_u from (2.129) into (2.127), we obtain

$$(k_1)_v = (k_2 - k_1)\frac{E_v}{2E}, \quad (k_2)_u = (k_1 - k_2)\frac{G_u}{2G}, \tag{2.130}$$

and Gauss's formula (2.126) takes the following form:

$$K = -\frac{1}{2\sqrt{EG}}\left[\left(\frac{E_v}{\sqrt{EG}}\right)_v + \left(\frac{G_u}{\sqrt{EG}}\right)_u\right]. \tag{2.131}$$

From the *Gauss–Peterson–Codazzi equations* follows the important *Bonnet rigidity theorem*.

Theorem 2.8.1 (O. Bonnet). *Let $E\,du^2 + 2F\,du\,dv + G\,dv^2$ and $L\,du^2 + 2M\,du\,dv + N\,dv^2$ be two arbitrary fundamental forms in a disk U of the (u, v)-plane, the first one positive definite. If for the coefficients of these fundamental forms the Gauss–Peterson–Codazzi equations are satisfied, then there is a unique, up to a rigid motion of the space $\mathbb{R}^3$, surface Φ given by $\vec{r}: U \rightarrow \mathbb{R}^3$ for which these forms are the first and the second fundamental forms, respectively.*

A proof of Bonnet's theorem can be found, for example, in [Kl2]. In the sense of ideas, this *fundamental theorem of surfaces* repeats the proof of Theorem 1.9.1, but in technical details it is lengthy enough and related with many computations.

2.8.3 Problems

Problem 2.8.1. If all the points of a regular complete surface Φ of class C^3 are umbilics, then Φ is a connected open region on either a sphere or a plane.

Solution. From the conditions of the problem it follows that the normal curvature $k(P, \vec{\lambda})$ of Φ at each point does not depend on $\vec{\lambda}$; hence, in particular, the principal curvatures $k_1(P)$ and $k_2(P)$ of the surface are equal to each other at every point P:

$$k_1(P) = k_2(P) = k(P). \tag{2.132}$$

Let P be an arbitrary point on Φ. Introduce an orthogonal coordinate system (u, v) in some neighborhood of P. Then for such a coordinate system, $F = 0$ holds, and since any curve on Φ is a line of curvature, $M = 0$ also holds. Then from (2.130) and (2.132) we obtain $(k)_u = (k)_v = 0$, i.e., in this case the function $k(P, \vec{\lambda})$ does not depend on either P or $\vec{\lambda}$; and now the assertion of the problem follows from Problem 2.4.1.   □

Problem 2.8.2. If the principal curvatures of a regular surface Φ of class C^2 are constants, then Φ is a sphere, a plane, or a cylinder.

Solution. Let $k_1(P) = c_1, k_2(P) = c_2$. If $c_1 = c_2$, then as was just proved, Φ is either a sphere, or, in the case of $c_1 = c_2 = 0$, a plane. Thus it is left to prove only that if $c_1 \neq c_2$, then either $c_1 = 0$ or $c_2 = 0$. Assume that the direction of a normal $\vec{n}$ on Φ is selected with the condition $k_2(P) > 0$, and suppose that $c_1 < c_2$. Let P be an arbitrary point on Φ. Introduce in a neighborhood of P a coordinate system (u, v) such that the coordinate curves are the lines of curvature. In this case, the Peterson–Codazzi equations take the form of (2.130). From the assumptions $k_1(P) = c_1, k_2(P) = c_2$, and $c_2 - c_1 \neq 0$ it follows that $E_v \equiv G_u \equiv 0$. Then Gauss's formula gives us

$$k_1 \cdot k_2 = K = -\frac{1}{2EG} (E_{vv} + G_{uu}) = 0. \tag{2.133}$$

From this follows $k_1 = 0$. The problem is solved. □

Problem 2.8.3. Find all complete regular surfaces of class C^3 without umbilics whose largest principal curvature is constant and equal to $\frac{1}{a}$.

Solution. In a neighborhood of an arbitrary point $P \in \Phi$ introduce a coordinate system (u, v) whose coordinate curves coincide with the lines of curvature. The Peterson–Codazzi equations take the form (2.130). From them and the condition of the problem it follows that

$$G_u \equiv 0, \tag{2.134}$$

but then from (2.93) it follows that a curve $\sigma_u : u = c, v = t$ is a geodesic. In fact, it is enough to prove that the value A in (2.93) that is equal to Γ_{22}^1 is zero. From (2.118) and (2.134) we have

$$\Gamma_{22}^1 = -\frac{G_u}{E} = 0. \tag{2.135}$$

So the curves σ_u are actually the lines of curvature, and at the same time the geodesics. Consequently, they are all plane curves, and their curvature $k_2(t)$ coincides with the principal curvature of Φ in the directions of the tangent lines to $\sigma_u(t)$; i.e., $k_u(t) = \frac{1}{a}$ holds. Hence $\sigma_u(t)$ is a circle C_u of radius a. Define now a curve γ_u as the geometrical locus of centers of the circles C_u. Thus, Φ is a *generalized cylinder* or a *generalized torus*; see Problem 2.8.2. □

Now consider surfaces whose principal curvatures are dependent by the relation

$$h(k_1(P), k_2(P)) = 0.$$

If h is symmetric with respect to k_1 and k_2, then Φ in this case is called a *Weingarten surface*. We consider the two most interesting and important cases[9] of h, namely, $k_1 + k_2$ and $k_1 \cdot k_2$.

---

[9] Classical examples of the case $k_1 + k_2$ are minimal surfaces and surfaces of constant mean curvature.

Theorem 2.8.2 (H. Liebmann). *If on a complete regular surface Φ of class C^3 the Gaussian curvature is constant and equal to some positive number $K_0 > 0$, then Φ is a sphere of radius $\frac{1}{\sqrt{K_0}}$.*

Proof. If the principal curvatures of Φ are equal to each other at all points of Φ, then the statement of Theorem 2.8.2 follows from Problem 2.8.1. Thus assume that there is a point P at which one of the principal curvatures, k_1, takes its maximum. Then the other principal curvature, k_2, takes its minimum at the same point; and this assumption leads to a contradiction. Since at P we have $k_1(P) \neq k_2(P)$, then in some neighborhood it is possible to introduce the coordinates (u, v) such that the coordinate curves will be the lines of curvature, or, what is the same, the equalities $M = F = 0$ hold. Moreover, at the origin $P(0, 0)$ the equations

$$(k_1)_u = (k_1)_v = (k_2)_u = (k_2)_v = 0 \tag{2.136}$$

and two inequalities

$$\frac{\partial^2 k_1}{\partial v^2} \leq 0, \quad \frac{\partial^2 k_2}{\partial u^2} \geq 0 \tag{2.137}$$

hold. We shall now use the Peterson–Codazzi equations in the form of (2.130). From them and from (2.136) we obtain

$$E_v(P) = G_u(P) = 0. \tag{2.138}$$

From (2.137) and (2.130) we have

$$\left[\frac{E_{vv}}{2E} (k_2 - k_1) \right]_P = \frac{\partial^2 k_1}{\partial v^2}(P) \leq 0,$$

and hence

$$E_{vv}(P) \geq 0. \tag{2.139}$$

Analogously,

$$\left[\frac{G_{uu}}{2G} (k_1 - k_2) \right]_P = \frac{\partial^2 k_2}{\partial u^2}(P) \geq 0, \tag{2.140}$$

from which follows

$$G_{uu}(P) \geq 0. \tag{2.141}$$

We substitute (2.138), (2.139), and (2.141) in Gauss's formula (2.131), and we obtain

$$K(P) = -\left[\frac{1}{2EG} (E_{vv} + G_{uu}) \right]_P \leq 0,$$

contrary to our assumption. This contradiction completes the proof of the theorem. $\qquad\square$

The completeness condition of the surface Φ in Theorem 2.8.2 is not superfluous. It is known that if the Gaussian curvature of some surface Φ is constant, then Φ is not necessarily part of a sphere. In other words, one can say that a complete sphere cannot be bent, but an arbitrary part of it can bend.

From Theorem 2.8.2 follows the following theorem of H. Liebmann.

Theorem 2.8.3. *If on a closed regular surface Φ the Gaussian curvature is positive and the mean curvature is constant and equal to H_0, then Φ is part of a sphere of radius $\frac{1}{H_0}$.*

Proof. A proof of the theorem can be given repeating scheme of the proof of Theorem 2.8.2 word for word.

Another proof of this theorem is based on Bonnet's method. Let $\Phi(a)$ be a parallel surface for Φ. Then by Theorem 2.4.6 the principal curvatures $k_1(a)$ and $k_2(a)$ of $\Phi(a)$ are expressed by the following formulas:

$$k_1(a) = \frac{k_1}{1 - ak_1}, \quad k_2(a) = \frac{k_2}{1 - ak_2}.$$

Hence the Gaussian curvature $K(a)$ of the surface $\Phi(a)$ is

$$K(a) = \frac{k_1 k_2}{1 - a(k_1 + k_2) + a^2 k_1 k_2} = \frac{k_1 k_2}{1 - 2H_0 a + a^2 k_1 k_2}.$$

Suppose that $a = \frac{1}{2H_0}$. Then $K \cdot \frac{1}{2H_0} = \frac{1}{4H_0^2}$. So we see that the Gaussian curvature of the surface $\Phi(\frac{1}{2H_0})$ is constant and equal to $\frac{1}{4H_0^2}$. Consequently, as was proved in Theorem 2.8.2, $\Phi(\frac{1}{2H_0})$ is a sphere. But then Φ is also a sphere of radius $\frac{1}{2H_0}$. $\square$

The condition of positivity of Gaussian curvature of a surface Φ in Theorem 2.8.3 is not superfluous, because if one of the principal curvatures at some point P of Φ is zero, then the other principal curvature at the same point is $2H_0$, and the point $\varphi(P)$ on $\Phi(\frac{1}{2H_0})$ is not regular, as is seen from Theorem 2.4.6.

Finally, if we require in the conditions of Theorem 2.8.3 that the sum of the principal curvature radii be constant, then the result would be the same. The surface Φ would be a sphere. This statement follows from a more general theorem by E.B. Christoffel, and it can be proved following the scheme of the proof of Theorem 2.8.2.

Finally, we shall prove Hilbert's theorem, as promised earlier.

Theorem 2.8.4 (D. Hilbert). *There does not exist a complete regular surface of class C^3 in $\mathbb{R}^3$ whose Gaussian curvature is a negative constant K_0.*

Proof. We assume the existence of a surface $\Phi \subset \mathbb{R}^3$ satisfying the conditions of the theorem and obtain a contradiction. Take on a surface a coordinate system (u, v) such that the coordinate curves are the lines of curvature. Then in the notation of Theorem 2.8.2 (Liebmann), we obtain the system of equations, see (2.130),

$$(k_1)_v = (k_2 - k_1)\frac{E_v}{2E}, \quad (k_2)_u = (k_1 - k_2)\frac{G_u}{2G}. \tag{2.142}$$

Assuming $K_0 = -1$ and integrating system (2.142), we obtain

$$E = \frac{C_1(u)}{1 + k_1^2(u, v)}, \quad G = \frac{C_2(v)}{1 + k_2^2(u, v)}.$$

The constants of integration $C_1(u)$, $C_2(v)$, without loss a generality, may be assumed to be 1. For this purpose it is enough to require that $E(u, 0) = \frac{1}{1+k_1^2(u,0)}$ and $G(0, v) = \frac{1}{1+k_2^2(0,v)}$, which one can always achieve. Thus, we obtain the equality

$$E = \frac{1}{1 + k_1^2}, \quad G = \frac{1}{1 + k_2^2}.$$

But then, in view of $k_1 k_2 = -1$,

$$E + G = \frac{1}{1 + k_1^2} + \frac{1}{1 + k_2^2} = \frac{2 + k_1^2 + k_2^2}{1 + k_1^2 + k_2^2 + k_1^2 k_2^2} = \frac{2 + k_1^2 + k_2^2}{2 + k_1^2 + k_2^2} = 1,$$

and hence we obtain

$$\sqrt{E} = \sin \sigma, \quad \sqrt{G} = \cos \sigma.$$

From this, using (2.128), we find that

$$L = + \sin \sigma \cos \sigma, \quad N = - \sin \sigma \cos \sigma.$$

Hence, a differential equation of asymptotic curves takes the form

$$(du + dv)(du - dv) = 0.$$

If we introduce new parameters p and q by the equalities

$$u = p - q, \quad v = p + q,$$

then the new coordinate curves $p = $ const and $q = $ const will be the asymptotic curves of the surface. The linear element of the surface will thus be presented by the following expression:

$$ds^2 = du^2 \sin^2 \sigma + dv^2 \cos^2 \sigma = dp^2 + 2 dp\, dq \cos \sigma + dq^2.$$

From here, by the way, one may conclude that the opposite sides in a quadrangle formed from asymptotic curves are equal. Such nets of curves (for which $E = G = 1$) on arbitrary surfaces were investigated by the Russian mathematician P.L. Tchebyshev in 1878. Such figures are formed by pulling a fishnet over a bent surface.

If we apply Gauss's formula (2.126) to the linear element expressed through parameters p, q, then for an angle $2\sigma = \omega$ formed by two asymptotic curves, we obtain the differential equation

$$\frac{\partial^2 \omega}{\partial p\, \partial q} = \sin \omega. \tag{2.143}$$

From here it is easy to deduce some simple corollaries; for example, for the area of a surface,

$$F = \iint \sin \omega \, dp \, dq.$$

For the area of a quadrangle bounded by asymptotic curves $p_1 < p < p_2$, $q_1 < q < q_2$, we obtain

$$F = \omega(p_1, q_1) - \omega(p_1, q_2) + \omega(p_2, q_2) - \omega(p_2, q_1),$$

or if we denote the inner angles by α_k (see Figure 2.20 a),

$$F = \alpha_1 + \alpha_2 + \alpha_3 + \alpha_4 - 2\pi \quad (0 < \alpha_k < \pi). \tag{2.144}$$

This formula was first given by J.N. Hazzidakis.[10] All known surfaces of constant negative curvature, as for example a helicoid found by F. Minding, have singular lines. Therefore, Hilbert asked whether there exist unbounded and in any finite area everywhere regular analytic surfaces with curvature $K = -1$. It turned out that such surfaces cannot exist.

If such a surface existed, then its asymptotic curves would be in any finite area everywhere regular analytic curves; two asymptotic curves would pass through each of its points with two different tangent lines, so that we could subordinate an angle between them to the condition $0 < \omega < \pi$. If we now accept that p and q are rectangular coordinates of a point in a plane, then we can put the points of our surface in one-to-one correspondence with the points of the plane.

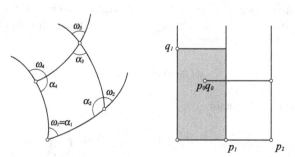

Figure 2.20. Tchebyshev net.

---

[10] *Über einige Eigenschaften der Flächen mit konstantem Krümmungsmasz*, Crelles J., v. 88, 68–73, 1880.

A surface will thus be mapped to the plane injectively. It is impossible to say in advance whether the inverse map will be also injective, since we did not yet exclude the possibility of the existence of closed asymptotic curves. Since by (2.143), ω cannot be constant on any of the asymptotic curves, then it is possible to select on the surface a reference mark, p and q, and a positive direction for counting p, so that $\omega(p, 0)$ will increase for

$$0 \le p \le p_2.$$

Then we obtain

$$\omega(p, q) - \omega(0, q) = \omega(p, 0) - \omega(0, 0) + \int_0^p \int_0^q \sin \omega \, dp \, dq. \qquad (2.145)$$

From this it follows that ω increases on each line segment $q = \text{const} > 0, 0 < p < p_2$, at least as fast as on the line segment $q = 0, 0 < p < p_2$ (since the double integral is positive).

Now consider a quadrangle (see Figure 2.20 b) $0 < p < p_1 < p_2, 0 < q < q_1$, and suppose $\omega(p_2, 0) - \omega(p_1, 0) = \varepsilon$. In this quadrangle, for sufficiently large q_1, obviously, there obviously exists a point for which

$$\omega = \pi - \frac{\varepsilon}{2}.$$

In fact, if ω always remains smaller than $\pi - \frac{\varepsilon}{2}$, then for sufficiently large q, the integral in (2.145) could be made as large as possible, since, for example, for

$$\frac{p_1}{2} < q < p_1$$

the inequalities

$$\omega\left(\frac{p_1}{2}, 0\right) - \omega(0, 0) < \omega(p, q) < \pi - \frac{\varepsilon}{2}$$

always hold, and hence $\sin \omega$ surpasses some positive number. But then it would be possible to make $\omega(p, q)$ large enough, contrary to the assumption that $\omega < \pi$. Now let (p_0, q_0) be a point such that

$$\omega(p_0, q_0) < \pi - \frac{\varepsilon}{2}.$$

In view of (2.145), we obtain

$$\omega(p_2, q_0) - \omega(p_0, q_0) = \omega(p_2, 0) - \omega(p_0, 0) + \int_{p_0}^{p_2} \int_0^{q_0} \sin \omega \, dp \, dq$$

$$> \omega(p_2, 0) - \omega(p_1, 0) = \varepsilon,$$

and hence $\omega(p_2, q_0) > \pi - \frac{\varepsilon}{2}$. Thus the angle would be greater than π on the line segment $p_0 < p < p_2; q = q_0$, in contradiction to our assumption. $\square$

This simple proof was given by E. Holmgren.[11]

Finally, we give in brief an idea of the first proof by Hilbert. From that fact that the asymptotic curves form a Tchebyshev net, Hilbert concludes that on a surface there is no closed asymptotic curve. Therefore, if a surface admits an everywhere "regular" net, then it is in one-to-one correspondence with a plane. From Hazzidakis's formula (2.144) it is possible to deduce as a corollary that the area of each quadrangle formed by the asymptotic curves is less than 2π and that hence the area of the entire surface,

$$\iint_{y>0} \frac{dx\,dy}{y^2}$$

is infinite. So, assuming that there is an unbounded and everywhere regular surface with measure of curvature $K = -1$, we come to a contradiction.

It is necessary to note only that the existence of the needed coordinate system on the entire surface was not proved.

2.9 Appendix: Indicatrix of a Surface of Revolution

Consider now the case of an arbitrary surface of revolution Φ, see Section 2.6.1. From the formulas

$$k_1 = \frac{f'}{x\sqrt{1+(f')^2}}, \quad k_2 = \frac{f''}{\left[1+(f')^2\right]^{3/2}}$$

it is clear that k_1 and k_2 depend on the same parameter. Hence, generally speaking, k_1 and k_2 are functionally dependent: $h(k_1, k_2) = 0$ or $k_2 = \varphi(k_1)$. Surfaces for which k_1 and k_2 are functionally related belong to the class of *Weingarten surfaces*; see Section 2.8.3. Not all of these surfaces are surfaces of revolution. The first question arising here is,

For what functions $k_2 = \varphi(k_1)$ do there exist complete surfaces of revolution?

The full answer to this question is given below without a proof in the case that the function φ is a bijection; that is, if $k_1(P) = k_1(Q)$ then $k_2(P) = k_2(Q)$ and conversely; here $P, Q \in \Phi$.[12]

Definition 2.9.1. The *indicatrix of a surface of revolution* Φ is the set of all pairs (k_1, k_2) counted at all points of the surface.

Theorem 2.9.1. *The indicatrix of a surface of revolution whose principal curvatures are related by a bijective functional dependence either reduces to a point or is the graph of a continuous strictly monotonic function defined on some real line segment.*

---

[11] Comptes Rendus, v. 134, 740–743, 1902.

[12] The results of Section 2.9 are due to V.V. Ivanov and were conveyed to the author in a personal communication.

Point-type indicatrices are of three kinds, a plane, a cylinder, or a sphere, by a suitable choice of a scale of measurements and a side of a surface (Figure 2.21). In each of these cases the corresponding complete surface is uniquely defined.

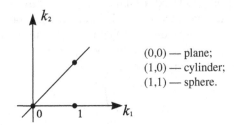

(0,0) — plane;
(1,0) — cylinder;
(1,1) — sphere.

Figure 2.21. Point-type indicatrices

(This statement coincides with Problem 2.8.2). Further, in speaking about the uniqueness of this or that surface of revolution we shall mean uniqueness to within the choice of an axis of rotation and a translation along it.

Theorem 2.9.2. *(a) An* increasing *graph serves as indicatrix of an accordingly oriented complete surface of revolution if and only if it belongs to one of the classes 1–6 described below.*

(b) A decreasing *graph serves as indicatrix of an accordingly oriented complete surface of revolution if and only if it belongs to one of the classes 7–9 described below.*

The proof of all these statements is long; it is based on the problem's reduction to studying of the following differential equation:

$$\frac{dv}{dx} = \varphi\left(\frac{v}{x}\right), \tag{2.146}$$

where $v = xk_1$. In what follows we only describe the above classes 1–9 of surfaces.

2.9.1 Increasing Indicatrices

In all six cases 1–6 described in this section, the indicatrix is the graph of a strictly increasing continuous and bounded function $k_2 = \varphi(k_1)$ defined on a finite line segment with endpoints a and b, always located over the diagonal $k_1 = k_2$ and resting on it by its left end.

Class 1.   The domain of the function φ is a line segment $a < k_1 < b$, where $0 < a < b$. Thus $\varphi(a) = a$ and $\varphi(k_1) > k_1$ when $a < k_1 < b$. For each function φ possessing the specified properties (Figure 2.22 a) there is a unique complete surface of revolution (Figure 2.22 b). This surface is closed, convex, and possesses a plane of symmetry orthogonal to the axis of rotation. Details of its shape depend on convergence of the integral

$$I = \int_a^b \frac{du}{\varphi(u) - u},$$

having a unique singularity at the point a. If $I = \infty$, then both curvatures decrease as they tend to the axis of rotation. In the case of $I < \infty$, spherical caps appear on the surface (Figure 2.22 c).

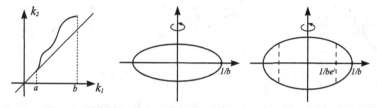

Figure 2.22. Increasing indicatrices: Class 1.

Class 2. A function φ is defined also on the line segment $a \le k_1 \le b$, but now $0 = a < b$. Moreover, $\varphi(0) = 0$ and $\varphi(k_1) > k_1$ when $0 < k_1 < b$. For each function φ possessing the specified properties (Figure 2.23 a) there is a unique complete surface of revolution (Figure 2.23 b). It is also closed, convex, and symmetric in the above-mentioned sense. For $I = \infty$ its shape does not differ from that specified in Figure 2.22 a, in this case only umbilics on the axis of rotation turn to planar points. If $I < \infty$, then instead of a spherical cap we see on the surface a flat circular region (Figure 2.23 c) whose radius decreases as the integral I increases.

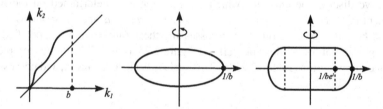

Figure 2.23. Increasing indicatrices: Class 2.

Class 3. The line segment $a \le k_1 \le b$ for which the function φ is defined now contains both positive and negative numbers: $a < 0 < b$. As before, $\varphi(a) = a$ and $\varphi(k_1) > k_1$ if $a < k_1 \le b$. In order that the graph (Figure 2.24 a) should serve as indicatrix of a complete surface of revolution, it is necessary and sufficient to fulfill an additional condition, namely, that the unique solution $v = v(x)$ (Figure 2.24 b) of the differential equation (2.146) that is determined on the interval $0 < x < 1/b$ and takes the value of 1 at $x = 1/b$ should satisfy the inequalities

$$v > -1, \qquad \int_0^{1/b} \frac{v\,dx}{\sqrt{1 - v^2}} > 0.$$

If desired, these requirements can be expressed directly in terms of the function φ.

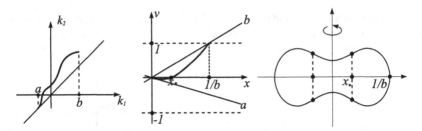

Figure 2.24. Increasing indicatrices: Class 3.

A complete surface of revolution in the case under discussion is also uniquely determined by its indicatrix. It is closed and symmetric, but completely loses its convexity, becoming similar to a sphere with "extended cheeks" (Figure 2.24 c). The presence or absence of an umbilical cap — only here concave — is determined by the above convergence condition of the integral I. As one may note, the three considered classes of bijective functional dependences between the principal curvatures and corresponding surfaces of revolution belong naturally to united series: the right endpoint of the indicatrix is located in the region of positive curvature, and the left endpoint passes from positive values through zero into the region of negative curvatures.

Class 4. The following three classes also form (in some sense) a separate family within the bounds of which a smooth transition from one case to other is possible. First, we discuss the class for which the function φ is determined on the line segment $a \leq k_1 \leq b$, where $a < b < 0$, and $\varphi(a) = a$ and $k_1 < \varphi(k_2) < 0$ if $a < k_1 \leq b$. For each function φ possessing these (and also those specified in the beginning of the section) properties (Figure 2.25 a) there is a unique complete surface of revolution (Figure 2.25 b). This surface is closed, convex, and possesses

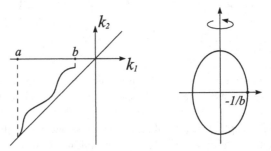

Figure 2.25. Increasing indicatrices: Class 4.

a plane of symmetry orthogonal to the axis of rotation. But, in contrast to surfaces of Class **1**, the curvatures now decrease with increasing distance from the axis of rotation: If the surfaces of the Class **1** were similar to a flattened ellipsoid, the new surfaces resemble an ellipsoid stretched along the axis of rotation. Note that in this case the previous remarks concerning the influence of the integral I on details of the shape of the surface near the axis of rotation remain valid.

Class 5.   The first case of the lack of uniqueness arises when the function φ, defined on the same line segment $a \le k_1 \le b$, $(a < b < 0)$, satisfies almost the same conditions: $\varphi(a) = a$ and $k_1 < \varphi(k_1) < 0$ if $a < k_1 < b$, but now only $\varphi(b) = 0$. The graph of any such function (Figure 2.26 a) is an indicatrix of several complete surfaces of revolution. All of them are convex, but only one of them is closed and symmetric, like a surface of the previous class. The only difference here is that our new surface comes to a point maximally far from the axis of rotation (incidentally, lying on the plane of symmetry), with zero curvature along a meridian. For this reason, its further behavior is not so rigidly limited, as before. The surface need not be reflected in the above-mentioned plane; it can

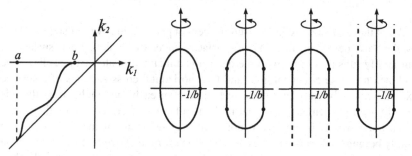

Figure 2.26. Increasing indicatrices: Class 5.

remain a cylinder for a while, even forever. Thus, we have four variants of a surface's behavior for a given indicatrix of the class under discussion; all of them are shown in Figures 2.26 b–e.

Class 6.   The last case of increasing indicatrix is one more example of the lack of uniqueness, but of absolutely another sort in comparison with the previous one. Here the function φ is determined on the line segment $a \le k_1 < 0$ and satisfies the conditions $\varphi(a) = a$ and $k_1 < \varphi(k_1) < 0$, but also, at the point 0 it has a zero left-hand limit.

For the existence of a complete surface of revolution whose indicatrix is the graph of such a function φ, it is necessary and sufficient that for some $u_* < 0$ the integral

$$\int_{u_*}^{0} \frac{\varphi(u)}{\varphi(u) - u} e^{I(u)} \, du \quad \text{where} \quad I(u) = \int_{u_*}^{u} \frac{dt}{\varphi(t) - t}$$

should converge. If one does not consider exotic (though quite possible) details of the behavior of the function φ at zero, this condition has a simple sense. For example, it is not satisfied for a function, whose graph has a nonzero inclination at the coordinate center. If $\varphi(u) = o(u)$, then the above condition is equivalent to the integrability of the function $\varphi(u)/u^2$ at zero.

The graph of the function satisfying the above-mentioned requirement (Figure 2.27 a) is an indicatrix of the entire family of complete surfaces of revolution (Figure 2.27 b).

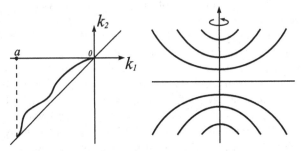

Figure 2.27. Increasing indicatrices: Class 6.

All of them are convex and are reminiscent of paraboloids by their shape, which by the way, belong to the class of surfaces under discussion. Each surface of our family runs to infinity with the determined inclination along a meridian. The inclination can be either infinite (and thus horizontal cross-sections of a surface extend without bound), or a nonzero number. For each such inclination, there is exactly one representative member in the family of surfaces corresponding to the given indicatrix. It is curious to note that if at least one of the surfaces in the family belongs to a cap (the presence of which for the above reasons depends on the nature of the approach of the left end of the indicatrix to the diagonal), then all the others also belong.

2.9.2 Decreasing Indicatrices

Quite a different picture is obtained, when an indicatrix is a decreasing line. In the remaining examples it is the graph of the function φ that (besides what will be told about it in each concrete case) is defined on the bounded line segment with endpoints $a < b \leq 0$, always including the point a, and is continuous and strictly decreasing on it. Thus the graph is located entirely above the umbilical diagonal, sometimes intersecting it at one point.

Class 7. The domain of the function $k_2 = \varphi(k_1)$ here is a line segment $a \leq k_1 < 0$, so on it $\varphi > 0$ holds. Thus the left-hand limit of the function φ at the point 0 should be zero. Any such function determines a unique complete saddle surface of revolution, homeomorphic to a cylinder with symmetric convex profile curve having a catenoid-like shape. But some important details depend on the asymptotic character of the function φ at zero. To see this, consider the integrals

$$I(u) = \int_a^u \frac{dt}{\varphi(t) - t} \quad \text{and} \quad J = \int_{-1/a}^0 \frac{v\,dx}{\sqrt{1 - v^2}}.$$

The first of them is an increasing function on the line segment $a \le u < 0$, and its limit for $u \to -0$ will be denoted by the letter I. In the second integral, which also could be expressed explicitly through φ, the function $v = v(x)$ means a unique solution of the differential equation (2.146), already familiar to us, existing on the interval $-1/a \le x < \infty$ and satisfying the initial condition $v(-1/a) = -1$. We have three cases.

Case 1. Let $I < \infty$ and so $J < \infty$. Then if one completes the graph of a function φ by one new point, attached to the origin (Figure 2.28 a), it serves as the indicatrix of an interesting surface whose profile curve is shown in Figure 2.28 b.

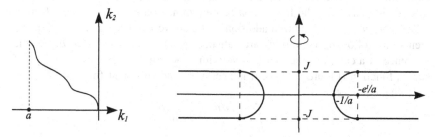

Figure 2.28. Decreasing indicatrices: Class 7, Case 1.

Case 2. Let $I = \infty$, but still $J < \infty$. In this case the indicatrix has no right endpoint (Figure 2.29 a), and the surface is similar to the previous one, and differs from it in that it no longer sticks to two infinite rings compressing it, though it asymptotically tends to them (Figure 2.29 b).

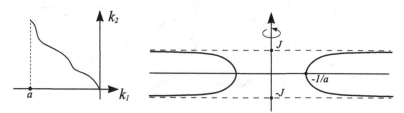

Figure 2.29. Decreasing indicatrices: Class 7, Case 2.

Case 3. Let finally $I = J = \infty$. This means that the indicatrix rapidly approaches the k_1-axis (Figure 2.30 a). In this case the distance between the branches of the profile curve (Figure 2.30 b) increases without bound while increasing its distance from the axis of rotation, but their inclination tends to a definite finite number. It is expressed by the formula

$$\frac{1-L}{\sqrt{L(2-L)}}, \quad \text{where} \quad L = -\frac{1}{a} \int_a^0 \frac{\varphi(u)}{\varphi(u) - u} e^{I(u)} \, du.$$

The values of the integral L fill the interval $0 < L \le 1$, so the asymptotic inclination of the branches can be any number, starting from zero.

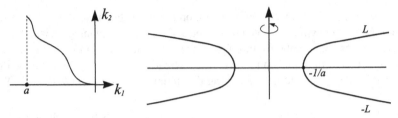

Figure 2.30. Decreasing indicatrices: Class 7, Case 3.

Class 8. A surface with a decreasing indicatrix can be periodic. Here we consider such a case. Let the function φ be determined on the line segment from a to b, where $a < b < 0$, and in addition to the above-mentioned general requirements, the following conditions are satisfied: $\varphi(a) > 0 > \varphi(b) > b$. For the existence of a complete surface of revolution whose indicatrix serves a graph of such a function (Figure 2.31 a), it is necessary and sufficient for it to obey the following equality:

$$I = \int_a^b \frac{du}{\varphi(u) - u} = \ln \frac{a}{b}.$$

Then the surface (Figure 2.31 b) together with all its parameters is uniquely defined by its indicatrix.

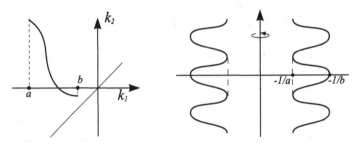

Figure 2.31. Decreasing indicatrices: Class 8.

Class 9. The last case we should discuss is similar to the previous one in many respects, but differs from it in the moment of principle: it provides one more example of the lack of uniqueness. Here the function φ is determined on the same line segment from a to b, where $a < b < 0$, but it now satisfies other conditions: $\varphi(a) > 0 > \varphi(b) = b$. The graph of such function a (Figure 2.32 a) serves as an indicatrix of a complete surface of revolution if and only if $I < \ln(a/b)$, where I is the integral just considered. But now this surface is already not unique (Figures 2.32 b – f). The reason is the following. The parallel farthest from the axis of rotation consists of umbilics, and in the case $I < \ln(a/b)$, it appears as an entire umbilic layer. But also its one circle is enough for that after its achievement a surface could select the variant of further behavior. At each such moment it has two variants — either to pass one more period repeating the previous form, or to curtail by a sphere. Thus there appears an infinite series of closed surfaces (Figures 2.32

b, c) similar to "pods of a peanut, containing some grains," and their number can be any natural number, but not smaller than two. If a surface once decides to stop, then we obtain two paraboloidal-type surfaces (i.e., homeomorphic to a plane) as shown in Figures 2.32 d, e. Finally, if a surface infinitely self-repeats the form, then it remains the same periodically pulsing cylindrical surface (Figure 2.32 f) that was seen in the previous class.

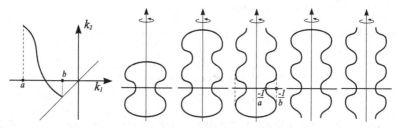

Figure 2.32. Decreasing indicatrices: Class 9.

2.10 Exercises to Chapter 2

Exercise 2.10.1. Write the parametric equations of (a) a twofold hyperbolic cylinder, (b) a circular cylinder, (c) a circular cone.

Exercise 2.10.2. Write the equations of a surface of revolution with the axis OZ:
(a) the torus obtained by a rotation of the circle $x = a + b \sin u$, $y = 0$, $z = b \sin u$, $0 < b < a$,
(b) the pseudosphere obtained by rotation of the *tractrix* $x = a \sin u$, $y = 0$, $z = a(\log \tan \frac{u}{2} + \cos u)$,
(c) the catenoid obtained by rotation of the *catenary* $x = a \cosh \frac{u}{a}$, $y = 0$, $z = u$.

Exercise 2.10.3. Show that the equations $x = \frac{u}{u^2+v^2}$, $y = \frac{v}{u^2+v^2}$, $z = \frac{1}{u^2+v^2}$ and $x = u \cos v$, $y = u \sin v$, $z = u^2$ define the same surface.

Exercise 2.10.4. Prove that the surface $x = \frac{a(uv+1)}{u+v}$, $y = \frac{b(u-v)}{u+v}$, $z = \frac{uv-1}{u+v}$ is a onefold hyperboloid. What are the coordinate curves of the surface in this parameterization?

Exercise 2.10.5. Given the surface $x = a(u + v)$, $y = b(u - v)$, $z = u + a$, find an explicit equation of the surface and prove that it is a hyperbolic paraboloid.

Exercise 2.10.6. Prove that the equation $x = u + \sin v$, $y = u + \cos v$, $z = u + a$ is the equation of a cylinder.

Exercise 2.10.7. Write the equations of the tangent plane and a normal to the surface (a) $x^2 + 2y^2 - 3z^2 - 4 = 0$ at the point $M(3, 1, -1)$, (b) $x = u + v$, $y = u - v$, $z = uv$ at the point $M(2, 1)$, (c) $x = u \cos v$, $y = u \sin v$, $z = av$ at each point.

Exercise 2.10.8. Write the equations of the tangent plane and a normal to the *pseudosphere*

$$x = a \sin u \cos v, \quad y = a \sin u \sin v, \quad z = a \left(\ln \tan \frac{u}{2} + \cos u \right).$$

Exercise 2.10.9. Show that all tangent planes to the surface $z = x^3 + y^3$ at the points $M(\alpha, -\alpha, 0)$ form a set of planes.

Exercise 2.10.10. Find the first fundamental form of (a) a sphere

$$\begin{cases} x = R \cos u \cos v, \\ y = R \cos u \sin v, \\ z = R \sin u, \end{cases}$$

(b) a circular cone $x = u \cos v$, $y = u \sin v$, $z = u$, (c) the *catenoid* $x = a \cosh \frac{u}{a}$, $y = u \cos \frac{u}{a} \sin v z = u$.

Exercise 2.10.11. Find the first fundamental form of each surface: (a) a cylinder, (b) a helicoid, (c) a pseudosphere.

Exercise 2.10.12. Let $v = \ln(u \pm \sqrt{u^2 + 1}) + C$ be the curves given on the *right helicoid* $x = u \cos v$, $y = u \sin v$, $z = 2v$. Calculate the arc lengths of these curves between the points $M_1(1, 2)$ and $M_1(3, 4)$.

Exercise 2.10.13. Find the area of the quadrangle on the *right helicoid* $x = u \cos v$, $y = u \sin v$, $z = av$ formed by the curves $u = 0$, $u = a$, $v = 0$, $v = 1$.

Exercise 2.10.14. Find the perimeter, inner angles, and the area of the curvilinear triangle
$u = \pm\frac{1}{2} avr$, $v = 1$ on the helicoid $x = u \cos v$, $y = u \sin v$, $z = -av$.

Exercise 2.10.15. Find the second fundamental form of each surface: (a) cylinder, (b) sphere, (c) circular cone, (d) helicoid, (e) catenoid, (f) *pseudosphere*

$$\begin{cases} x = a \sin u \cos v, \\ y = a \sin u \sin v, \\ z = a \left(\log \tan \frac{u}{2} + \cos u \right). \end{cases}$$

Exercise 2.10.16. Find the principal curvatures (a) at the vertices of the twofold hyperboloid $\frac{x^2}{a^2} - \frac{y^2}{b^2} - \frac{z^2}{c^2} - 1 = 0$, (b) of the surface $z = xy$ at the point $M(1, 1, 1)$.

Exercise 2.10.17. Find the principal curvatures and the principal vectors of each surface:

(a) $z = xy$ at the point $M(1, 1, 1)$,   (b) $\frac{x^2}{p} + \frac{y^2}{p} = 2z$ at the point $M(0, 0, 0)$,

(c) $x = u^2 + v^2$, $y = u^2 - v^2$, $z = uv$ at the point $M(1, 1)$.

Exercise 2.10.18. Find the Gaussian curvature of each surface: (a) a sphere, (b) a (circular) cylinder, (c) a pseudosphere, (d) a catenoid, (e) a paraboloid $\frac{x^2}{p} + \frac{y^2}{p} = 2z$, (f) a helicoid, (g) a torus.

Exercise 2.10.19. Calculate the mean curvature of a catenoid.

Exercise 2.10.20. Prove that the umbilics on the surface $x = \frac{u^2}{2} + v$, $y = u + \frac{v^2}{2}$, $z = uv$ lie on the curves $u = v$, $u + v + 1 = 0$.

Exercise 2.10.21. Find the lines of curvature on the following surfaces: (a) an arbitrary surface of revolution, (b) the surface $x = u^2 + v^2$, $y = u^2 - v^2$, $z = v$.

Exercise 2.10.22. Find the lines of curvature on each surface: (a) $x = u^2 + v^2$, $y = u^2 - v^2$, $z = v$, (b) $x = 3u^2 - u^3 + 3uv^2$, $y = u^2 - 3u^2v - 3v$, $z = 3(u^2 - v^2)$.

Exercise 2.10.23. Prove that the curve $x = \frac{2}{1+t}$, $y = \frac{2}{1-t}$, $z = t$ is an asymptotic curve on the surface $z = \frac{1}{x^2} - \frac{1}{y^2}$.

Exercise 2.10.24. Find the asymptotic curves of a right helicoid.

Exercise 2.10.25. Find the geodesic curvature of each curve:
  (a) a circle of radius r on a sphere of radius R,
  (b) a helix $u = c$, $v = t$ on a helicoid $x = u \cos v$, $y = u \sin v$, $z = u$.

Exercise 2.10.26. Find the geodesic curvature of the curves $u = $ const and $v = $ const on a surface $x = u \cos v$, $y = y \sin v$, $z = f(v)$.

Exercise 2.10.27. Prove that if no external forces act on a particle moving along a surface, then it will move along a geodesic.

Exercise 2.10.28. Prove that every straight line on a surface is a geodesic.

Exercise 2.10.29. Prove that a geodesic is also a line of curvature if and only if it is a straight line.

Exercise 2.10.30. Prove that a geodesic is also an asymptotic curve if and only if it is a plane curve.

Exercise 2.10.31. Two surfaces touch each other along a curve L. Prove that if L is a geodesic on one of the surfaces, then it is also a geodesic on the other surface.

Exercise 2.10.32. Find the geodesics of (1) a cylindrical surface, (2) a circular cone, (3) a *Möbius strip*, (4) a *flat torus*, (5) a *Klein bootle*.

Hint. The surfaces (4), (5) are obtained from a square in the plane with the standard metric after gluing the pair(s) of its sides with corresponding orientation.

Exercise 2.10.33. Prove that the meridians of a surface of revolution are its geodesics.

Exercise 2.10.34. Prove that a parallel of a surface of revolution is also its geodesic if and only if a tangent line to a meridian passing through a point is parallel to the axis of rotation.

Exercise 2.10.35. Find the geodesics on (a) a *right helicoid*: $x = u \cos v$, $y = u \sin v$, $z = u$, (b) a pseudosphere.

Exercise 2.10.36. Prove that the only minimal surfaces of revolution are a plane and a catenoid.

Exercise 2.10.37. Prove that among all ruled surfaces, only a plane and a catenoid are minimal.

Exercise 2.10.38.*Let γ be a closed geodesic without points of self-intersection on a closed convex surface. Prove that the spherical image of γ divides a sphere into two parts with equal areas.

3

Intrinsic Geometry of Surfaces

In this chapter the foundations of intrinsic geometry of a two-dimensional surface are introduced. The material is presented in such a form that all statements other than the Gauss–Bonnet theorem can be generalized almost word for word to the multidimensional case. So, this chapter can be considered as an *introduction to the theory of n-dimensional Riemannian manifolds.*

3.1 Introducing Notation

The local coordinates (u, v) of some parameterization of a surface Φ will be written as (u^1, u^2), the vectors $\vec{r}_{u^1}(u^1, u^2)$ and $\vec{r}_{u^2}(u^1, u^2)$ by $\vec{r}_1$ and $\vec{r}_2$. Denote the coefficients of the first fundamental form by g_{ik} $(i, k = 1, 2)$, where $g_{ik} = \langle \vec{r}_i, \vec{r}_k \rangle$; the elements of the inverse matrix to $\{g_{ij}\}$ by g^{ij}; the coordinates of the vector $\vec{\lambda} = \lambda^1 \vec{r}_1 + \lambda^2 \vec{r}_2$ in a local basis $\vec{r}_1$, $\vec{r}_2$ by λ^1, λ^2. Introduce the magnitudes, called the *Christoffel symbols of the first kind,*

$$\Gamma_{jk,l} = \frac{1}{2} \left(\frac{\partial g_{jl}}{\partial u^k} + \frac{\partial g_{kl}}{\partial u^j} - \frac{\partial g_{jk}}{\partial u^l} \right) \qquad (i, j, k = 1, 2), \qquad (3.1)$$

and the *Christoffel symbols of the second kind* by the formula

$$\Gamma^i_{jk} = \sum_{l=1}^{2} \Gamma_{jk,l} g^{li} \qquad (i, j, k = 1, 2). \qquad (3.2)$$

The derivational formulas (2.109) in this notation take the form

$$\frac{\partial \vec{r}_i}{\partial u^j} = \vec{r}_{ij} = \sum_{k=1}^{2} \Gamma_{ij}^k \vec{r}_k + A_{ij}\vec{n} \qquad (i, j = 1, 2), \qquad (3.3)$$

where $\vec{n}$ is a normal to Φ, and A_{ij} are the coefficients of the second fundamental form. The Gaussian curvature K of Φ is expressed through the coefficients of the first fundamental form by the formula, see (2.123),

$$K = \frac{1}{g_{11}} \left(\frac{\partial \Gamma_{11}^2}{\partial u^2} - \frac{\partial \Gamma_{11}^2}{\partial u^1} + \Gamma_{11}^1 \Gamma_{12}^2 + \Gamma_{11}^2 \Gamma_{22}^2 - \Gamma_{12}^1 \Gamma_{11}^2 - \Gamma_{12}^2 \Gamma_{12}^2 \right). \qquad (3.4)$$

If the equations of a curve $c(t)$ in local coordinates (u^1, u^2) are given by the functions $(u^1(t), u^2(t))$, then the tangent vector to $c(t)$, which is equal to $(u^1)'\vec{r}_1 + (u^2)'\vec{r}_2$, is denoted by $\dot{c}(t)$. Later on, we will agree that if in any term the same index is written twice, one time *up* and the second time *down*, then it in fact occurs as a summation with respect to this index. For example, $(\vec{\mu}, \vec{\lambda}) = g_{ij}\lambda^i \mu^j$, $\Gamma_{jk}^i = g^{il}\Gamma_{jk,l}$, and so on.

3.2 Covariant Derivative of a Vector Field

Let $\vec{\lambda}(t) \in T\Phi_{c(t)}$ be a differentiable vector field along a smooth curve $c(t)$ on a regular surface Φ of class C^3. Take the derivative $\frac{d}{dt}\vec{\lambda}$ of this vector field at a point $c(t)$. In the general case, $\frac{d}{dt}\vec{\lambda}$ does not belong to the tangent plane $T\Phi_{c(t)}$. Define the *covariant derivative* $\frac{D}{dt}\vec{\lambda}$ of the vector field $\vec{\lambda}$ along $c(t)$ as the orthogonal projection of $\frac{d}{dt}\vec{\lambda}$ onto $T\Phi_{c(t)}$. From this definition it follows that

$$\frac{D}{dt}\vec{\lambda} = \frac{d}{dt}\vec{\lambda} - \left\langle \frac{d}{dt}\vec{\lambda}, \vec{n} \right\rangle \vec{n}, \qquad (3.5)$$

where $\vec{n}$ is a normal to Φ at $c(t)$. To find an analytical expression for $\frac{D}{dt}\vec{\lambda}$ in local coordinates, let $\vec{\lambda}(t) = \lambda^1(t)\vec{r}_1(t) + \lambda^2(t)\vec{r}_2(t)$, and let $c(t)$ be given by the equations $u^1 = u^1(t)$, $u^2 = u^2(t)$. From this notation we obtain

$$\frac{d}{dt}\vec{\lambda} = \frac{d\lambda^1}{dt}\vec{r}_1 + \frac{d\lambda^2}{dt}\vec{r}_2 + \lambda^1 \left(\vec{r}_{11}\frac{du^1}{dt} + \vec{r}_{12}\frac{du^2}{dt} \right) + \lambda^2 \left(\vec{r}_{21}\frac{du^1}{dt} + \vec{r}_{22}\frac{du^2}{dt} \right).$$

We now express $\vec{r}_{1i}$ and $\vec{r}_{2j}$ through $\vec{r}_1, \vec{r}_2$ and $\vec{n}$ by derivational formulas (3.3), obtaining

$$\frac{d}{dt}\vec{\lambda} = \left(\frac{d\lambda^1}{dt} + \Gamma_{ij}^1 \lambda^i \frac{du^j}{dt} \right) \vec{r}_1 + \left(\frac{d\lambda^2}{dt} + \Gamma_{ij}^2 \lambda^i \frac{du^j}{dt} \right) \vec{r}_2 + A\vec{n}. \qquad (3.6)$$

From (3.6) it is seen that $\langle \frac{d}{dt}\vec{\lambda}, \vec{n} \rangle = A$. Therefore,

$$\frac{D}{dt}\vec{\lambda} = \left(\frac{d\lambda^1}{dt} + \Gamma_{ij}^1 \lambda^i \frac{du^j}{dt} \right) \vec{r}_1 + \left(\frac{d\lambda^2}{dt} + \Gamma_{ij}^2 \lambda^i \frac{du^j}{dt} \right) \vec{r}_2 \qquad (3.7)$$

and

$$\left(\frac{D}{dt}\vec{\lambda}\right)^k = \frac{d\lambda^k}{dt} + \Gamma_{ij}^k \lambda^i \frac{du^j}{dt} \qquad (k = 1, 2). \tag{3.8}$$

Remark 3.2.1. Sometimes it is useful to use $\frac{D}{dt}\vec{\lambda}|_{c(t)}$ instead of the notation $\frac{D}{dt}\vec{\lambda}$, indicating explicitly that the covariant derivative is taken along $c(t)$.

Remark 3.2.2. In the definition of covariant derivative of a vector field $\vec{\lambda}(t)$ along $c(t)$ we have used the circumstance that the surface Φ lies in $\mathbb{R}^3$. However, (3.7), (3.8), and (3.3) show that for a given curve $c(t)$ and the vector field $\vec{\lambda}(t)$ the coordinates of a vector field $\frac{D}{dt}\vec{\lambda}$ are expressed only through the coefficients of the first fundamental form, their first-order derivatives, and the coordinates of $\vec{\lambda}$. Thus one can say that the *covariant derivative of a vector field along a curve is an object of intrinsic geometry.*

3.3 Parallel Translation of a Vector along a Curve on a Surface

Definition 3.3.1. A vector field $\vec{\lambda}(t)$ along a curve $c(t)$ on a surface Φ is called a *field of parallel vectors* if $\frac{D}{dt}\vec{\lambda}(t)|c(t) \equiv 0$.

Definition 3.3.2. We say that a vector $\vec{\lambda}_1$ at a point $c(t_1)$ is obtained from a vector $\vec{\lambda}_0$ at a point $c(t_0)$ by a *parallel translation along* $c(t)$ $(t_0 \le t \le t_1)$ if there is a field of parallel vectors $\vec{\lambda}(t)$ $(t_0 \le t \le t_1)$ along $c(t)$ for which $\vec{\lambda}(t_0) = \vec{\lambda}_0$, $\vec{\lambda}(t_1) = \vec{\lambda}_1$.

If a curve $c(t)$ lies entirely in one coordinate neighborhood, then for expressing the coordinates λ_1^1, λ_1^2 of a vector $\vec{\lambda}_1$ through coordinates λ_0^1, λ_0^2 of a vector $\vec{\lambda}_0$, in view of Definition 3.3.2, we need to solve the system, see (3.8),

$$\frac{d\lambda^k}{dt} + \Gamma_{ij}^k \lambda^i \frac{du^j}{dt} = 0 \qquad (k = 1, 2) \tag{3.9}$$

with initial conditions $\lambda^1(t_0) = \lambda_0^1, \lambda^2(t_0) = \lambda_0^2$ and to set $\lambda_1^1 = \lambda^1(t_1), \lambda_1^2 = \lambda^2(t_1)$. The existence and uniqueness of a solution of the system (3.9) for arbitrary initial conditions and for any t follow from a theorem about systems of ordinary differential equations.

If $c(t)$ does not lie entirely in one coordinate neighborhood, then we divide it onto a finite number of arcs, each of which belongs to some coordinate neighborhood, and then produce a parallel translation along each part in consecutive order. Parallel translation of a vector can also be defined along a piecewise smooth curve as consecutive parallel translations along each smooth part. Note that a parallel displacement of a vector on a surface is defined as a parallel translation along a given curve. Thus, in general, the result of a parallel translation depends on the curve along which it is considered. But if there is a region B on Φ for which the result of a parallel displacement does not depend on the choice of curve, but depends only on its endpoints, then one says that there is *absolute parallelness on the region B*.

Obviously, the parallel translation together with the covariant derivative are objects of the intrinsic geometry. This circumstance allows us to produce the parallel displacement in simple cases without writing and solving the system (3.9).

Example 3.3.1. Take a cone with vertex at the point O and with a solid angle φ (Figure 3.1). Let c be a curve connecting the points P_0 and P_1, and let $\vec{\lambda}_0$ be any vector at P_0. Find a vector $\vec{\lambda}_1$ obtained from $\vec{\lambda}_0$ by a parallel translation along c. Since a cone is locally isometric to a plane, we act in the following way: "*cut*" a cone along a straight line, its ruling, and develop it onto the Euclidean plane $\mathbb{R}^2$. Obtain on the plane a region inside of an angle. Construct a vector $\vec{\lambda}_1$ at the point P_1 that is parallel to $\vec{\lambda}_0$ in the sense of parallel translation in $\mathbb{R}^2$. If $\vec{\lambda}_0$ forms an angle α with a ruling OP_0, then a vector $\vec{\lambda}_1$ will form an angle $\alpha + \beta$ with a ruling OP_1, where $\beta = \angle P_0 O P_1$. We thus see that on any simply connected region B on a cone without the vertex O there is absolute parallelness. However, under a parallel translation along a closed curve that envelops the vertex O, a vector rotates through an angle β; thus absolute parallelness does not exist on the cone as a whole.

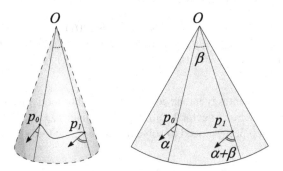

Figure 3.1. Parallel translation along a cone and its envelope.

3.3.1 Properties of the Parallel Translation

Theorem 3.3.1. *Parallel translation of vectors along a curve preserves the scalar product of vectors and linear operations with them.*

Proof. The second statement of the theorem follows directly from the definition, and also from the linearity of the system (3.9). We prove the first statement. Let $\vec{\lambda}_1(t)$ and $\vec{\lambda}_2(t)$ be two fields of parallel vectors along some curve $c(t)$. We must show that $\langle \vec{\lambda}_1(t), \vec{\lambda}_2(t) \rangle \equiv$ const. This statement follows from the formula

$$\frac{d}{dt}\left\langle \vec{\lambda}_1, \vec{\lambda}_2 \right\rangle = \left\langle \frac{D}{dt}\vec{\lambda}_1, \vec{\lambda}_2 \right\rangle + \left\langle \vec{\lambda}_1, \frac{D}{dt}\vec{\lambda}_2 \right\rangle. \tag{3.10}$$

Let us prove it:

$$\frac{d}{dt}\langle \vec{\lambda}_1, \vec{\lambda}_2 \rangle = \left\langle \frac{d}{dt}\vec{\lambda}_1, \vec{\lambda}_2 \right\rangle + \left\langle \vec{\lambda}_1, \frac{d}{dt}\vec{\lambda}_2 \right\rangle.$$

From the definition of covariant derivative it follows that $\left\langle \frac{d}{dt}\vec{\lambda}_1, \vec{\lambda}_2 \right\rangle = \left\langle \frac{D}{dt}\vec{\lambda}_1, \vec{\lambda}_2 \right\rangle$ and $\left\langle \vec{\lambda}_1, \frac{d}{dt}\vec{\lambda}_2 \right\rangle = \left\langle \vec{\lambda}_1, \frac{D}{dt}\vec{\lambda}_2 \right\rangle$, from which follows (3.10). The statement of the theorem may be also proven by direct use of (3.9):

$$\langle \vec{\lambda}_1, \vec{\lambda}_2 \rangle = g_{ij}\lambda_1^i \lambda_2^j,$$

$$\frac{d}{dt}\left(g_{ij}\lambda_1^i \lambda_2^j \right) = \frac{\partial g_{ij}}{\partial u^k}\frac{du^k}{dt}\lambda_1^i \lambda_2^j + g_{ij}\frac{d\lambda_1^i}{dt}\lambda_2^j + g_{ij}\lambda_1^i \frac{d\lambda_2^j}{dt}.$$

Substitute here $\frac{d\lambda_1^i}{dt}$ and $\frac{d\lambda_2^j}{dt}$ from (3.9), and then use (3.7) and (3.8). We obtain

$$
\begin{aligned}
\frac{d}{dt}\langle \vec{\lambda}_1, \vec{\lambda}_2 \rangle &= \frac{\partial g_{ij}}{\partial u^k}\frac{du^k}{dt}\lambda_1^i \lambda_2^j - g_{ij}\Gamma^i_{pk}\lambda_1^p \frac{du^k}{dt}\lambda_2^j - g_{ij}\Gamma^j_{pk}\lambda_2^p\frac{du^k}{dt}\lambda_1^i \\
&= \frac{\partial g_{ij}}{\partial u^k}\frac{du^k}{dt}\lambda_1^i \lambda_2^j - \Gamma_{pk,j}\lambda_1^p \lambda_2^j \frac{du^k}{dt} - \Gamma_{pk,i}\lambda_2^p \lambda_1^i \frac{du^k}{dt} \\
&= \frac{\partial g_{ij}}{\partial u^k}\lambda_1^i \lambda_2^j \frac{du^k}{dt} - \frac{1}{2}\left(\frac{\partial g_{pi}}{\partial u^k} + \frac{\partial g_{kj}}{\partial u^p} - \frac{\partial g_{pk}}{\partial u^j} \right)\lambda_1^p \lambda_2^j \frac{du^k}{dt} \\
&\quad - \frac{1}{2}\left(\frac{\partial g_{pi}}{\partial u^k} + \frac{\partial g_{kj}}{\partial u^p} - \frac{\partial g_{pk}}{\partial u^j} \right)\lambda_1^i \lambda_2^p \frac{du^k}{dt} \\
&= \left[\frac{\partial g_{ij}}{\partial u^k} - \frac{1}{2}\left(\frac{\partial g_{ij}}{\partial u^k} + \frac{\partial g_{kj}}{\partial u^i} - \frac{\partial g_{ik}}{\partial u^j} \right) - \frac{1}{2}\left(\frac{\partial g_{ij}}{\partial u^k} + \frac{\partial g_{ki}}{\partial u^j} - \frac{\partial g_{jk}}{\partial u^i} \right) \right] \\
&\quad \cdot \lambda_1^i \lambda_2^j \frac{du^k}{dt} = 0.
\end{aligned}
$$

Thus based on the notion of covariant derivative of a vector field we can define the notion of a parallel translation of a vector along a curve.

Conversely, we can define the covariant derivative of a vector field from the notion of parallel translation of a vector along a curve. Indeed, let $c(t)$ be a curve and $\vec{\lambda}(t)$ a vector field along $c(t)$. Take two vectors $\vec{\lambda}(t)$ and $\tilde{\lambda}(t + \Delta t)$ at a point $c(t)$ and assume that $\tilde{\lambda}(t + \Delta t)$ is obtained from $\vec{\lambda}(t + \Delta t)$ by a parallel translation along the arc $c(t + \Delta t)c(t)$ of a curve $c(t)$. Then define

$$\frac{D}{dt}\vec{\lambda}\Big|_c = \lim_{\Delta t \to 0} \frac{\tilde{\lambda}(t + \Delta t) - \vec{\lambda}(t)}{\Delta t}. \tag{3.11}$$

Note that if a parallel translation of a vector in a local coordinates is defined by (3.9), then (3.8) can be deduced from (3.11). The proof of this statement is left to the reader as an exercise.　□

3.4 Geodesics

In this section we continue the study the geodesics. It turns out that the properties of geodesics imitate the properties of straight lines in a plane as far as possible. One of the characteristic properties of a straight line in a plane is that a vector tangent to it stays tangent under parallel translation. Define a geodesic on a surface Φ as a curve with the same property.

3.4.1 Definition of Geodesics and Their Equations

Let $\gamma(t)$ be a twice continuously differentiable curve on a regular surface Φ of class C^2.

Definition 3.4.1. A curve $\gamma(t)$ is called a *geodesic on a surface* Φ if the vector field $\dot{\gamma}(t)$ is a field of parallel vectors along $\gamma(t)$.

We deduce the equations of geodesics in local coordinates. If the equations of $\gamma(t)$ are $u^1 = u^1(t)$, $u^2 = u^2(t)$, then the vector $\dot{\gamma}(t)$ has coordinates $\frac{du^1}{dt}$, $\frac{du^2}{dt}$. Consequently, from (3.8) follows

$$\left(\frac{D}{dt}\dot{\gamma}\right)^k = \frac{d^2u^k}{dt^2} + \Gamma_{ij}^k \frac{du^i}{dt}\frac{du^j}{dt} \qquad (k = 1, 2),$$

and we obtain the *equations of geodesics*

$$\frac{d^2u^k}{dt^2} + \Gamma_{ij}^k \frac{du^i}{dt}\frac{du^j}{dt} = 0 \qquad (k = 1, 2). \tag{3.12}$$

Note that from our definition of geodesic it follows that the parameter t is the *canonical parameter*, i.e., it is proportional to the arc length. In fact, since $\dot{\gamma}(t)$ is a field of parallel vectors, then from Theorem 3.3.1 it follows that $|\dot{\gamma}(t)| = c$ and

$$s = \int_0^t |\dot{\gamma}(t)|\,dt = ct. \tag{3.13}$$

Example 3.4.1. We find all geodesics on a cylinder with one more method. As is seen from the definition of geodesics or directly from the system (3.12), the geodesics under an isometric (or locally isometric) map of one surface onto other also turn into geodesics. Use this property for finding the geodesics on a circular cylinder of radius R. Cutting the cylinder along its ruling and developing it onto a plane, we obtain in the plane a strip of width $2\pi R$ between parallel straight lines a_1 and a_2. However, it is more convenient to do the following. Denote by $\vec{d}$ a vector orthogonal to a_1 and of length $2\pi R$. Take an arbitrary point P inside the strip and identify with it all points in the plane that can be obtained by parallel displacement of P by a vector multiple of the vector $\vec{d}$. This identification allows us to build a local isometry of the plane onto the cylinder (a *covering map*). Under this map any straight line in the plane will turn into some geodesic on the cylinder, and all geodesics on the cylinder can be thus obtained. This circumstance allows us to find all geodesics on the cylinder. Straight lines that are parallel to a_1 turn into rulings of the cylinder; straight lines that are orthogonal to a_1 turn into closed geodesics (parallels on a cylinder); and straight lines that are oblique to a_1 turn into helices on the cylinder. Thus, the *geodesics on a cylinder are exactly the rulings, parallels, and helices.*

Problem 3.4.1. Study the behavior of geodesics on a circular cone.

We prove now that the definitions of geodesics given in Chapters 2 and 3 are equivalent. Let $\gamma(s)$ be a length-parameterized curve. Then $\frac{D}{ds}\dot{\gamma} = 0$ means that $\frac{d}{ds}\dot{\gamma} = \vec{v}$ is a vector parallel to $\vec{n}$, and conversely, if the vector $\frac{d}{ds}\dot{\gamma}$ is parallel to $\vec{n}$, then $\frac{D}{ds}\dot{\gamma} = 0$.

3.4.2 Exponential Map. Properties of an Exponential Map and Local Properties of Geodesics

The system (3.12) is a system of second-order ordinary differential equations, solved with respect to the highest derivatives. Therefore, the next lemma follows from well-known theorems on the existence and uniqueness of solutions of systems of ordinary differential equations and the continuous dependence of their solutions on initial values.

Lemma 3.4.1. For each point P on a surface Φ there are a neighborhood U of P and a real number $\tilde{\epsilon}_P > 0$ such that for every point $Q \in U$ and an arbitrary vector $\vec{\lambda} \in T\Phi_Q$ whose length is smaller than $\tilde{\epsilon}_P$ there is a unique geodesic $\gamma(Q, \vec{\lambda}, t)$ $(-1 < t < 1)$ satisfying the conditions $\gamma(Q, \vec{\lambda}, 0) = Q$, $\dot{\gamma}(Q, \vec{\lambda}, 0) = \vec{\lambda}$.

From our notation and from (3.13) follows the equality

$$\gamma(Q, t \cdot \vec{\lambda}, 1) = \gamma(Q, \vec{\lambda}, t). \tag{3.14}$$

For some point Q and some vector $\vec{\lambda} \in T\Phi_Q$ let there be a geodesic $\gamma(Q, \vec{\lambda}, t)$ $(0 \le t \le 1)$. Denote by $\exp_Q\vec{\lambda}$ the point $\gamma(Q, \vec{\lambda}, 1)$:

$$\exp_Q\vec{\lambda} = \gamma(Q, \vec{\lambda}, 1). \tag{3.15}$$

In view of Lemma 3.4.1, for each point $Q \in \Phi$ and each vector $\vec{\lambda} \in T\Phi_Q$ whose length is not greater than $\tilde{\epsilon}_Q$ there is a geodesic $\gamma(Q, \vec{\lambda}, t)$ for $0 \le t \le 1$. Consequently, (3.15) defines some map from an open disk of radius $\tilde{\epsilon}_Q$ with center at Q in the plane $T\Phi_Q$ into Φ. This mapping is called an *exponential map*.

Problem 3.4.2. Describe the exponential map of a tangent plane on a sphere of radius R. Find all critical points of this map.

Lemma 3.4.2. The rank of the exponential map $\exp_Q\vec{\lambda}$ at the point Q is 2.

Proof. Let the equations of geodesics $\gamma(Q, \vec{\lambda}, t)$ in local coordinates u^1, u^2 be

$$u^1 = f^1(\lambda^1, \lambda^2, t), \qquad u^2 = f^2(\lambda^1, \lambda^2, t).$$

Then the mapping $\exp_Q\vec{\lambda}$ is given by the functions $f^1(\lambda^1, \lambda^2, 1)$, $f^2(\lambda^1, \lambda^2, 1)$. By the definition of a geodesic $\gamma(Q, \vec{\lambda}, t)$,

$$\left.\frac{df^i}{dt}\right|_{t=0} = \lambda^i \qquad (i = 1, 2). \tag{3.16}$$

But in view of (3.14), the following identity holds:

$$f^i(\lambda^1, \lambda^2, t) = f^i(t\lambda^1, t\lambda^2, 1) \qquad (i = 1, 2). \tag{3.17}$$

Differentiating the identity (3.17) by t and assuming $t = 0$, we obtain

$$\left.\frac{df^i}{dt}(\lambda^1, \lambda^2, t)\right|_{t=0} = \frac{\partial f^i(0, 0, 1)}{\partial \lambda^1}\lambda^1 + \frac{\partial f^i(0, 0, 1)}{\partial \lambda^2}\lambda^2 \qquad (i = 1, 2). \tag{3.18}$$

Now from (3.16) and (3.18) it follows that the Jacobian matrix of the mapping $\exp_Q$ at the point Q is the unit matrix. $\qquad\square$

From Lemmas 3.4.1 and 3.4.2 we shall deduce the main lemma of this section.

Lemma 3.4.3. For each point $Q \in \Phi$ there is a neighborhood W_Q and a real ϵ_Q such that for any two points $Q_1 \in W_Q$ and $Q_2 \in W_Q$ there is a unique geodesic $\gamma(Q_1, Q_2, t)$ $(0 \le t \le 1)$ joining Q_1 and Q_2 and of length not greater than ϵ_Q.

Proof. Let U_Q and $\tilde{\epsilon}_Q$ be the neighborhood and real number obtained in Lemma 3.4.1. Denote by $A(U_Q, \tilde{\epsilon}_Q)$ the set of pairs $(R, \vec{\lambda})$, where $R \in U_Q$ and $\vec{\lambda} \in T\Phi_R$. Introduce coordinates on a set A by corresponding to each element $(R, \vec{\lambda})$ of the set $A(U_Q, \tilde{\epsilon}_Q)$ the real numbers u^1, u^2, λ^1, λ^2, where u^1, u^2 are local coordinates of the point R, and λ^1, λ^2 are the coordinates of $\vec{\lambda}$ in a local basis of the plane $T\Phi_R$. Then define the set $B = U_Q \times U_Q$. Also introduce coordinates in the set B by corresponding to each element (Q_1, Q_2) of the set B the real numbers x^1, x^2, y^1, y^2, where x^1, x^2 are local coordinates of the point Q_1, and y^1, y^2 are local coordinates of the point Q_2. Define a map $\mathbf{f}: A \to B$ in the following way: correspond the pair $(R, \vec{\lambda})$ of A to the pair $(R, \exp_R\vec{\lambda})$ of B.

(α). Prove that the rank of the map $\mathbf{f}$ at the point $(Q, 0)$ is maximal, i.e., equal to 4. By the notation of the previous lemma, the map $\mathbf{f}$ can be rewritten in the following form:

$$x^1 = u^1, \quad x^2 = u^2, \quad y^1 = f^1(u^1, u^2, \lambda^1, \lambda^2, 1),$$
$$y^2 = f^2(u^1, u^2, \lambda^1, \lambda^2, 1). \tag{3.19}$$

From (3.19) it is seen that the value of the determinant of the Jacobi matrix of the map $\mathbf{f}$ at the point Q is equal to the value of the determinant $\begin{vmatrix} \partial f^1/\partial \lambda^1 & \partial f^1/\partial \lambda^2 \\ \partial f^2/\partial \lambda^1 & \partial f^2/\partial \lambda^2 \end{vmatrix}$ derived for $\lambda^1 = \lambda^2 = 0$. The value 1 of the last determinant was calculated in Lemma 3.4.2. The statement (α) is proven.

Continuing, we deduce from the inverse function theorem the existence of a neighborhood W_Q and a real ϵ_Q such that the map $\mathbf{f}$ is a diffeomorphism of the set $A(W_Q, \epsilon_Q)$ onto B. Let $C = \mathbf{f}(A)$. Take a neighborhood W_Q of the point Q such that $W_Q \times W_Q \subset C$. Then the neighborhood W_Q and ϵ_Q satisfy all the requirements of the lemma. $\qquad\square$

By the way, we have proved that the coordinates of the point $\gamma(Q_1, Q_2, t)$ and the function $l(Q_1, Q_2)$, the length of the geodesic $\gamma(Q_1, Q_2, t)$, smoothly depend on the coordinates of the points Q_1 and Q_2 for $Q_1 \neq Q_2$.

Example 3.4.2. For a sphere S_R^2 of radius R in $\mathbb{R}^3$, any circular neighborhood W_P of radius ϵ with center at an arbitrary point P has all the properties given in Lemma 3.4.3 if $\epsilon < \frac{1}{2}\pi R$. Indeed, recall that the geodesics on a sphere are the arcs of great circles. If $P_1 \in W_P$ and $P_2 \in W_P$, then for $\epsilon < \frac{1}{2}\pi R$ the points P_1, P_2 and the center O of the sphere lie on the same straight line. Pass a plane through them. This plane intersects the sphere in a great circle, which, by the construction, is uniquely defined by the points P_1 and P_2, and the shortest arc of this circle lies entirely in W_P and its length is smaller than ϵ.

Problem 3.4.3. Prove that for a circular cylinder of radius R any circular neighborhood W_P of radius ϵ with center at the point P has all the properties from Lemma 3.4.3 if $\epsilon < \frac{1}{2}\pi R$.

3.4.3 Parallel Translation and Geodesic Curvature of a Curve

The geodesic curvature k_g of an arbitrary curve on a surface was defined in Section 2.7.3. Now we shall give another (equivalent) definition of k_g using the notion of parallel translation of a vector along a curve. Let $c(t)$ be a parameterization of a curve c, and t an arc length parameter. Take two vectors $\dot{c}(t)$ and $\tilde{c}'(t + \Delta t)$ at a point $c(t)$, where the vector $\tilde{c}'(t + \Delta t)$ is obtained from $\dot{c}(t + \Delta t)$ by parallel translation along the arc $c(t + \Delta t)c(t)$ of the curve c. Denote by $\Delta\psi(\Delta t, t)$ the angle between $\dot{c}(t)$ and $\tilde{c}'(t + \Delta t)$. Define now the *geodesic curvature* by the formula

$$k_g = \lim_{\Delta t \to 0} \frac{\Delta\psi(\Delta t, t)}{\Delta t}. \tag{3.20}$$

Theorem 3.4.1. *For any regular curve $c(t)$ of class C^2, the geodesic curvature exists and is*

$$k_g = \left| \frac{D}{dt}\dot{c} \right|, \tag{3.21}$$

where t is an arc length parameter.

Proof. Since $\dot{c}(t)$ and $\tilde{c}'(t + \Delta t)$ are unit vectors, then $2 \sin \frac{\Delta \psi}{2} = |\tilde{c}'(t + \Delta t) - \dot{c}(t)|$. Hence

$$k_g = \lim_{\Delta t \to 0} \frac{\Delta \psi}{\Delta t} = \lim_{\Delta t \to 0} \frac{\Delta \psi}{2 \sin \frac{\Delta \psi}{2}} \cdot \lim_{\Delta t \to 0} \frac{|\tilde{c}'(t + \Delta t) - \dot{c}(t)|}{\Delta t} = \left| \frac{D}{dt} \dot{c} \right|$$

by formula (3.11).

We prove now that (3.21) and (2.88) of Chapter 2 coincide. Indeed, from the definition of covariant derivative of a vector field, $\frac{D}{dt}\dot{c}(t)$ is the projection of the vector $\frac{d}{dt}\dot{c}(t)$ on the tangent plane $T\Phi_{c(t)}$. Therefore,

$$\left| \frac{D}{dt} \dot{c}(t) \right| = \left| \frac{d}{dt} \dot{c}(t) \right| \sin \varphi,$$

where φ is the angle between the main normal to the curve $c(t)$ and the normal to the surface, but $\left| \frac{d}{dt}\dot{c} \right| = k$, and we have $\left| \frac{D}{dt}\dot{c} \right| = k \sin \varphi$, which completes the proof.   □

From Theorem 3.4.1 it follows that if $k_g = 0$ at each point of $c(t)$, then $c(t)$ is a geodesic.

For curves on a surface Φ one can define the *sign of geodesic curvature* similarly, as was done for the curvature of plane curves. Recall one particular case that will be important for us in what follows. Let c be a regular curve bounding a region D homeomorphic to a disk. If a vector $\frac{D}{dt}\dot{c}$ at some point is directed inside of D, then assume that the geodesic curvature of c at this point is positive, and in the opposite case, negative.

Example 3.4.3. Consider the intersection of the sphere $x^2 + y^2 + z^2 = R^2$ with the plane $z = a$ $(-R < a < R)$. This intersection is the circle S_a on the sphere. The radius of this circle in the plane is $\sqrt{R^2 - a^2}$. Hence the curvature of S_a is $1/\sqrt{R^2 - a^2}$, and the vector $\frac{d}{dt}\dot{c}$ is directed inside of the circle. The angle that $\frac{d}{dt}\dot{c}$ forms with the tangent plane to the sphere is equal to $\arccos \frac{a}{R}$. Consequently, the geodesic curvature with respect to the region $D : z \leq a$, depending on the sign, is equal to $\frac{a}{R\sqrt{R^2-a^2}}$. In particular, for $a = 0$ we obtain a great circle whose geodesic curvature is zero, i.e., a geodesic on the sphere.

Problem 3.4.4. Find the geodesic curvature of the intersection of the paraboloid $z = x^2 + y^2$ with the plane $z = a^2$. The sign of geodesic curvature should be determined with respect to the region $0 \leq z \leq a^2$.

3.4.4 Geodesics and Parallel Translation

Parallel translation along geodesics on a surface has an especially simple realization. Indeed, the tangent vector to a geodesic stays tangent during parallel translation along it, and parallel translation of vectors keeps their scalar product. Thus for drawing a field of parallel vectors along a geodesic it is sufficient to build a vector of constant length at each point that forms a constant angle with the tangent vector to the curve.

Example 3.4.4. Consider again a two-dimensional sphere in $\mathbb{R}^3$. Construct two meridians orthogonal to each other from the north pole O of the sphere. Denote by A_1 and A_2 their intersections with the equator. Consider a closed piecewise smooth curve γ composed of geodesic arcs $\sigma_1 = OA_1$, $\sigma_2 = A_1A_2$, and $\sigma_3 = A_2O$. Let $\vec{\lambda}$ be a unit vector tangent to σ_1 at the point O. Find the vector $\vec{\lambda}_1$ obtained from $\vec{\lambda}$ by parallel translation along σ_1. Since σ_1 is a geodesic arc, then $\vec{\lambda}_1$ is a vector of the same length and again tangent to σ_1, but just at the point A_1. Furthermore, $\vec{\lambda}_1$ is orthogonal to σ_2. Consequently, $\vec{\lambda}_2$ obtained from $\vec{\lambda}_1$ by a parallel translation along σ_2 is again a vector orthogonal to σ_2, but just at the point A_2. Further, $\vec{\lambda}_1$ is tangent to σ_3, but σ_3 is a geodesic, thus a vector $\vec{\lambda}_3$ obtained from $\vec{\lambda}_2$ by parallel translation along σ_3 is also tangent to σ_3, but at O. Finally, we see that $\vec{\lambda}$ turns through the angle $\pi/2$ under parallel translation along γ.

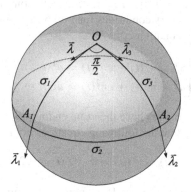

Figure 3.2. Parallel translation along geodesics on a sphere.

3.5 Shortest Paths and Geodesics

3.5.1 Metric on a Surface and the Shortest Paths

Recall the definition of the distance between two points on a surface Φ. Let P and Q be two points on Φ. Denote by $\mathbf{L}_{PQ}$ the *set of all curves on the surface* Φ *with endpoints at* P *and* Q. The *distance* $\rho_\Phi(P, Q)$ between the points P and Q is defined by the formula

$$\rho_\Phi(P, Q) = \inf_{c \in \mathbf{L}_{PQ}} l(c). \tag{3.22}$$

It is not difficult to show that on a connected differentiable surface Φ the distance is well-defined for any pair of points. To prove this, it is sufficient to show that any two points on such a surface can be joined by a curve of finite length. The function $\rho_\Phi \geq 0$ has all the usual properties of a metric (see Section 2.3.2):

$$
\begin{align*}
(1) \quad & \rho_\Phi(P, Q) = \rho_\Phi(Q, P), \\
(2) \quad & \rho_\Phi(P, Q) + \rho_\Phi(Q, R) \geq \rho_\Phi(P, R), \tag{3.23} \\
(3) \quad & \rho_\Phi(P, Q) = 0 \iff P = Q.
\end{align*}
$$

Further, in the cases in which it would not lead to a contradiction, the distance $\rho_\Phi(P, Q)$ between the points P and Q is denoted by PQ.

Definition 3.5.1. The *shortest path joining two points P and Q* is a curve, whose length is PQ.

3.5.2 Stationary Curves of the Length Functional

Let γ be a twice continuously differentiable curve on Φ located in some coordinate neighborhood U; $u^i = u^i(t)$ $(i = 1, 2)$ its equations, t a canonical parameter, $a \leq t \leq b$. Find necessary conditions for γ to be a shortest path. Consider the curve $\gamma_\varepsilon(t)$ given by the equations

$$
\begin{cases}
u^1 = u^1(t) + \epsilon\eta^1(t) = \tilde{u}^1(t, \varepsilon), \\
u^2 = u^2(t) + \varepsilon\eta^2(t) = \tilde{u}^2(t, \varepsilon),
\end{cases} \tag{3.24}
$$

and require that the curve $\gamma_\varepsilon(t)$ have the same endpoints as a curve $\gamma(t)$. For this it is sufficient to assume that the following equations hold:

$$
\eta^i(a) = \eta^i(b) = 0 \qquad (i = 1, 2). \tag{3.25}
$$

From (3.24) it follows that $\gamma_0 = \gamma$. If we now fix the functions η^1 and η^2, then $l(\varepsilon) = l(\gamma_\varepsilon)$, and the length of γ_ε is a function of ε only. Thus for γ to be a shortest path, it is necessary that it satisfy

$$
\begin{cases}
\dfrac{dl}{d\varepsilon}\bigg|_{\varepsilon=0} = 0 \\
l(\varepsilon) = \displaystyle\int_a^b F(\tilde{u}^i, \tilde{u}^{i'})\, dt,
\end{cases} \tag{3.26}
$$

where

$$
\tilde{u}^{i'} = \frac{d\tilde{u}^i}{dt}, \qquad F(\tilde{u}^i, \tilde{u}^{i'}) = \sqrt{g_{ij}(\tilde{u}^i)\tilde{u}^{i'}\tilde{u}^{j'}},
$$

$$
\frac{dl}{d\varepsilon}\bigg|_{\varepsilon=0} = \int_a^b \left(\frac{\partial F}{\partial \tilde{u}^i}\eta^i + \frac{\partial F}{\partial \tilde{u}^{i'}}\eta^{i'} \right) dt = \frac{\partial F}{\partial \tilde{u}^i}\eta^i \bigg|_a^b - \int_a^b \left(\frac{\partial F}{\partial \tilde{u}^i} - \frac{d}{dt}\frac{\partial F}{\partial \tilde{u}^{i'}} \right) \eta^i\, dt.
$$

The nonintegral term is zero in view of the conditions (3.25). Thus from (3.26) we obtain

$$
\int_a^b \left(\frac{\partial F}{\partial u^i} - \frac{d}{dt}\frac{\partial F}{\partial u^{i'}} \right) \eta^i\, dt = 0. \tag{3.27}
$$

Recall now that $\eta^1(t)$ and $\eta^1(t)$ are arbitrary functions satisfying the conditions (3.25) only. Set

$$f^1 = \frac{\partial F}{\partial u^1} - \frac{d}{dt}\frac{\partial F}{\partial u^{1\prime}}, \qquad f^2 = \frac{\partial F}{\partial u^2} - \frac{d}{dt}\frac{\partial F}{\partial u^{2\prime}},$$

and assume that

$$\eta^1 = f^1 \sin^2 \frac{\pi(t-a)}{b-a}, \qquad \eta^2 = f^2 \sin^2 \frac{\pi(t-a)}{b-a}.$$

Substituting $\eta^1(t)$ and $\eta^2(t)$ into (3.27), we obtain

$$\int_a^b [(f^1)^2 + (f^2)^2] \sin^2 \frac{\pi(t-a)}{b-a} \, dt = 0,$$

from which follows $f^1 = f^2 = 0$, or

$$\frac{d}{dt}\frac{\partial F}{\partial u^i} - \frac{\partial F}{\partial u^{i\prime}} = 0 \qquad (i = 1, 2). \tag{3.28}$$

The system of equations (3.28) is called an *Euler system*, and the curves that are the solutions of this system are called the *stationary curves of the length functional*. Writing down the equations (3.28) for the length functional $F = \sqrt{g_{ij}u^{i\prime}u^{j\prime}}$, we have

$$\frac{\partial F}{\partial u^{i\prime}} = \frac{g_{pi}u^{p\prime}}{\sqrt{g_{ij}u^{i\prime}u^{j\prime}}} = \frac{1}{C}g_{pi}u^{p\prime},$$

and because t is a canonical parameter, we have $\sqrt{g_{ij}u^{i\prime}u^{j\prime}} = C$. Therefore,

$$\frac{d}{dt}\left(\frac{\partial F}{\partial u^{i\prime}}\right) = \frac{1}{C}\left(\frac{\partial g_{pk}}{\partial u^i}\frac{du^p}{dt}\frac{du^k}{dt} + g_{pi}\frac{d^2u^p}{dt^2}\right).$$

Moreover,

$$\frac{\partial F}{\partial u^i} = \frac{1}{2\sqrt{g_{ij}u^{i\prime}u^{j\prime}}}\left(\frac{\partial g_{pk}}{\partial u^i}\frac{du^p}{dt}\frac{du^k}{dt}\right) = \frac{1}{2C}\left(\frac{\partial g_{pk}}{\partial u^i}\frac{du^p}{dt}\frac{du^k}{dt}\right).$$

Substituting the expressions for $\frac{\partial F}{\partial u^i}$ and $\frac{d}{dt}\left(\frac{\partial F}{\partial u^{i\prime}}\right)$ in (3.28), we obtain

$$g_{ip}\frac{d^2u^p}{dt^2} + \frac{\partial g_{pi}}{\partial u^k}\frac{du^k}{dt}\frac{du^p}{dt} - \frac{1}{2}\frac{\partial g_{pk}}{\partial u^i}\frac{du^p}{dt}\frac{du^k}{dt} = 0. \tag{3.29}$$

Note that

$$\frac{\partial g_{pi}}{\partial u^k}\frac{du^k}{dt}\frac{du^p}{dt} = \frac{1}{2}\left(\frac{\partial g_{pi}}{\partial u^k} + \frac{\partial g_{ki}}{\partial u^p}\right)\frac{du^p}{dt}\frac{du^k}{dt}.$$

Therefore, (3.29) can be rewritten in the following form:

$$g_{ip}\frac{d^2u^p}{dt^2} + \frac{1}{2}\left(\frac{\partial g_{pi}}{\partial u^k} + \frac{\partial g_{ki}}{\partial u^p} - \frac{\partial g_{pk}}{\partial u^i}\right)\frac{du^p}{dt}\frac{du^k}{dt} = 0 \qquad (i = 1, 2),$$

or, using expressions for Christoffel symbols of the first kind,

$$g_{ip}\frac{d^2u^p}{dt^2} + \Gamma_{pk,i}\frac{du^p}{dt}\frac{du^k}{dt} = 0 \qquad (i = 1, 2). \tag{3.30}$$

Multiply (3.30) by g^{ji} and sum it over i. Then (3.2) gives us in final form,

$$\frac{d^2u^j}{dt^2} + \Gamma^j_{pk}\frac{du^p}{dt}\frac{du^k}{dt} = 0 \qquad (j = 1, 2). \tag{3.31}$$

Thus, we see that stationary curves of the length functional coincide with geodesics, as they were defined in Section 3.4.

It is necessary to note that from this result it does not follow that any geodesic is a shortest path, because only necessary conditions for the minimum of the length functional were found. Moreover, it is easy to give an example when an arc of a geodesic is not a shortest path. For this, it is sufficient to take an arc of a great circle on a sphere whose length is greater than the semiperimeter of the whole circle. Also it does not follow from our result that a shortest path is a geodesic, because we have assumed in our deducing of equations (3.31) that a shortest path γ is twice continuously differentiable curve, which does not follow from the above text.

3.5.3 Geodesics as Shortest Paths

Here we prove that sufficiently small arcs of geodesics are shortest paths. Hence the curves realizing the distance between their endpoints, i.e., the shortest paths, are sometimes called *segments* or *minimal geodesics*.

Let P be a point on Φ and let W_P be a neighborhood in which any two points can be joined by a unique geodesic with length not greater than ε_P (see Lemma 3.4.3). Suppose that two length-parameterized curves $\gamma_1(s)$ and $\gamma_2(s)$ are given in the neighborhood W_P. Join the points $\gamma_1(s)$ and $\gamma_2(s)$ by the geodesic $\gamma(s, t) = \gamma(\gamma_1(s), \gamma_2(s), t)$, where t is a canonical parameter, counting from the point $\gamma_1(s)$. Denote by $\alpha(s)$ the angle between $\dot{\gamma}_1$ and $-\dot{\gamma}$ at the point $\gamma_1(s)$, and by $\beta(s)$ the angle between $\dot{\gamma}_2$ and $\dot{\gamma}$ at a point $\gamma_2(s)$. The function $l(s) = l(\gamma(s, t))$ is differentiable with respect to s, in view of the remark to Lemma 3.4.3.

Lemma 3.5.1 (The derivative of the length of a family of geodesics).

$$\frac{dl}{ds} = \cos\alpha(s) + \cos\beta(s). \tag{3.32}$$

Proof. Let the equations of the curves $\gamma_1(s)$ and $\gamma_2(s)$ be presented by the functions $f^i_k(s)$; let $u^i = f^i_k(s)$ $(k = 1, 2)$, and let the equations of geodesics $\gamma(s, t)$ be the functions $\varphi^i(s, t)$. Then the functions $\varphi^i(s, t)$ satisfy the conditions

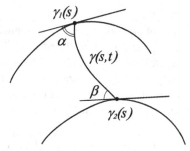

Figure 3.3. Derivative of the length of a family of geodesics.

$$\varphi^i(s, 0) = f_1^i(s), \quad \varphi^i(s, 1) = f_2^i(s) \qquad (i = 1, 2), \qquad (3.33)$$

$$\cos\alpha(s) = \frac{-g_{ij}\frac{df_1^i}{ds}\cdot\frac{d\varphi^j}{dt}(s, 0)}{\sqrt{g_{ij}\frac{d\varphi^i}{dt}(s, 0)\cdot\frac{d\varphi^j}{dt}(s, 0)}},$$

$$\cos\beta(s) = \frac{-g_{ij}\frac{df_2^i}{ds}\cdot\frac{d\varphi^j}{dt}(s, 1)}{\sqrt{g_{ij}\frac{d\varphi^i}{dt}(s, 1)\cdot\frac{d\varphi^j}{dt}(s, 1)}}.$$

(3.34)

Set $F = \sqrt{g_{ij}\dot\varphi^i\dot\varphi^j}$ and obtain

$$\frac{dl}{ds} = \int_0^1\left(\frac{\partial F}{\partial\dot\varphi^i}\cdot\frac{\partial^2\varphi^i}{\partial s\partial t} + \frac{\partial F}{\partial\varphi^i}\cdot\frac{\partial\varphi^i}{\partial s}\right)dt$$

$$= \frac{\partial F}{\partial\dot\varphi^i}\cdot\frac{\partial\varphi^i}{\partial s}\Big|_0^1 + \int_0^1\left(\frac{\partial F}{\partial\varphi^i} - \frac{d}{dt}\frac{\partial F}{\partial\dot\varphi^i}\right)dt.$$

The subintegral expression in parentheses is zero, because all the curves $\gamma(s, t)$ are geodesics. Thus

$$\frac{dl}{ds} = \frac{\partial\varphi^i}{\partial s}(s, 1)\cdot\frac{\partial F}{\partial\dot\varphi^i}\Big|_{t=1} - \frac{\partial\varphi^i}{\partial s}(s, 0)\cdot\frac{\partial F}{\partial\dot\varphi^i}\Big|_{t=0},$$

and since $\frac{\partial F}{\partial\dot\varphi^i} = \frac{g_{ij}\dot\varphi^i}{\sqrt{g_{ij}\dot\varphi^i\dot\varphi^j}}$, then from (3.33) and (3.34) we obtain (3.32).   □

Remark 3.5.1. If one of the curves, say γ_1, degenerates to a point, then $\frac{dl}{ds} = \cos\beta(s)$.

We now prove an important lemma.

Lemma 3.5.2. For each point P there is a neighborhood such that any two points from this neighborhood can be joined by a unique shortest path, and this shortest path is a geodesic.

Proof. Let W_P be a neighborhood defined in Lemma 3.4.3. Choose a neighborhood $\tilde V_P$ of the point P such that for any two points Q_1 and Q_2 from this neighborhood an arc $\gamma(Q_1, Q_2)$ of a geodesic connecting them lies in W_P.

(α). Prove that $l_0 = l(\gamma(Q_1, Q_2))$, the length of the geodesic $\gamma(Q_1, Q_2)$, is not greater than the length s_0 of any other curve $c(s)$ connecting the same points Q_1 and Q_2 and located in W_P; and $s_0 = l_0$ if and only if $c(s) \equiv \gamma$. Here s is the arc length of $c(s)$ counting from Q_1. Join Q_1 with a point $c(s)$ by a geodesic $\sigma(s)$. Denote by $l(s)$ the length of $\sigma(s)$ and apply to this function the remark of Lemma 3.5.1. Then $\frac{dl}{ds} = \cos\beta(s)$, where $\beta(s)$ is the angle between $\sigma(s, t)$ and $\dot{c}(s)$ at a point of their intersection. From this follows

$$l_0 = \int_0^{s_0} \frac{dl}{ds} ds = \int_0^{s_0} \cos\beta(s) ds \leq \int_0^{s_0} ds = s_0 = l(c(s)).$$

The statement (α) is completely proved.

Further, take a disk K_1 in a plane $T\Phi_P$ with center at a point P and with radius ε_1 so small that $V_P^1 = \exp_P(K_1) \subset \tilde{V}_P$ holds. Consider, finally, a circular neighborhood V_P^2 of radius $\frac{1}{4}\varepsilon_1$. Prove that a neighborhood V_P^2 can be taken in a part of a neighborhood V_P. Indeed, let Q_1 and Q_2 be arbitrary points in V_P and $\gamma(Q_1, Q_2)$ a geodesic joining them. Prove that $\gamma(Q_1, Q_2)$ is a shortest path. Let $c(t)$ be an arbitrary curve on Φ with endpoints Q_1 and Q_2. If this curve

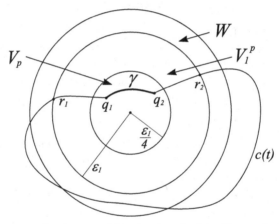

Figure 3.4. Any two points from a neighborhood can be joined by a unique shortest path.

lies entirely in W_P and does not coincide with $\gamma(Q_1, Q_2)$, then its length $l(c)$ is greater than the length of $\gamma(Q_1, Q_2)$ by the statement (α) just proven.

Therefore, assume that $c(t)$ does not lie entirely in W_P. Then it does not lie entirely in the neighborhood V_P^1. Denote by R_1 (R_2) the first point of intersection of $c(t)$ with the boundary of V_P^1 counting from Q_1 (from Q_2). The length of the radius PR_1 is equal to ε_1, the length of the radius PQ_1 is not greater than $\frac{1}{4}\varepsilon_1$, and the length of the curve composed from the radius PQ_1 and the arc Q_1R_1 of the curve $c(t)$, by statement (α), is not smaller than the length of PR_1. Consequently, the length QR_1 of the curve $c(t)$ is not smaller than $\varepsilon_1 - \frac{\varepsilon_1}{4} = \frac{3}{4}\varepsilon_1$. Analogously, the length of the arc R_2Q_2 of the curve $c(t)$ is not smaller than $\frac{3}{4}\varepsilon_1$. Hence the length of the whole curve $c(t)$ in the case under discussion is not smaller than

$\frac{3}{2}\varepsilon_1$. On the other hand, the length of the geodesic $\gamma(Q_1, Q_2)$, again in view of statement (α), is not greater than the sum of the lengths of radii PQ_1 and PQ_2, which is equal to $\frac{1}{2}\varepsilon_1$. Comparing these two results, we obtain the statement of the lemma.   □

A neighborhood where the statements of Lemma 3.5.2 hold is called *canonical*.

Problem 3.5.1. Prove that the neighborhoods mentioned in Example 3.4.2 and Problem 3.4.3 are canonical neighborhoods.

From Lemma 3.5.2 we can deduce some important theorems.

Theorem 3.5.1. *For each inner point P on a geodesic γ there is an arc QR such that $P \in QR$ and the arc QR is a shortest path.*

Proof. Let V_P be a canonical neighborhood of a point P. Since P is an inner point of γ, then there exist two points Q and R on the geodesic such that P belongs to the arc QR and both points lie inside of V_P. Then by Lemma 3.5.2, the arc QR is a shortest path.   □

The statement of Theorem 3.5.1 can also be reformulated in the following form: *any sufficiently small arc of a geodesic is a shortest path.*

Indeed, sufficiently large arcs of geodesics can also be shortest paths. For example, any arc of a geodesic with length not greater than πR on a sphere of radius R and on a circular cylinder of radius R is a shortest path.

The following theorem about convex surfaces was formulated by A.D. Aleksandrov and proved by A.V. Pogorelov [Pog].

Theorem 3.5.2. *On a convex surface whose Gaussian curvature is not greater than k_0, any arc of a geodesic of length not greater than $\pi/\sqrt{k_0}$ is a shortest path.*

Theorem 3.5.3. *Any shortest path is a geodesic.*

Proof. Let V_P be a canonical neighborhood of an arbitrary point P located on some shortest path γ. Take two arbitrary points $Q_1 \in \gamma$ and $Q_2 \in \gamma$ in V_P and join them by a geodesic $\gamma(Q_1, Q_2)$. By Lemma 3.5.2, this geodesic is the unique shortest path joining Q_1 and Q_2. Consequently, $\gamma(Q_1, Q_2) \subset \gamma$. Since Q_1, Q_2 were chosen arbitrarily, then our theorem is completely proved.   □

Theorem 3.5.4. *For each compact set F there is a real $d > 0$ such that any two points from F whose mutual distance is smaller than d, can be joined by a unique shortest path.*[13]

Proof. Let $\{V_P\}$ be a system of canonical neighborhoods that are constructed for all $P \in F$. This system covers the whole of F. Since F is a compact set, one can select a finite cover $V_1 = V_{P_1}, \ldots, V_n = V_{P_n}$. Consider a real d (*Lebesgue number*) so small that any two points whose mutual distance is smaller than d belong to one of the above neighborhoods $\{V_i\}_{1 \leq i \leq n}$. Then the statement of the theorem follows from Lemma 3.5.2.   □

---

[13] d is an *elementary length of the compact set F*.

Theorem 3.5.5. *If two different shortest paths γ_1 and γ_2 have two common points P_1 and P_2, then these points are the ends of γ_1, and so of γ_2.*

Proof. Assume the contrary. Let, for instance, P_2 be an inner point of γ_2. Denote by $\overline{P_1 P_2}$ the arc of a shortest path γ_1, and by $\overline{\overline{P_1 P_2}}$ the arc of a shortest path γ_2. Take a point $Q_1 \in \overline{P_1 P_2}$, $Q_1 \neq P_2$, in a canonical neighborhood V_{P_2} of P_2 and a point $Q_2 \neq P_2$, $Q_2 \in \gamma_2$, $Q_2 \notin \overline{\overline{P_1 P_2}}$. By the triangle inequality we have

$$Q_1 Q_2 \leq Q_1 P_2 + P_2 Q_2. \tag{3.35}$$

But by Lemma 3.5.2, the equality in (3.35) is possible if and only if the arc $Q_1 P_2$ lies on γ_2, and from this it would follow by Theorem 3.5.3 that the shortest paths γ_1 and γ_2 coincide. Hence

$$Q_1 Q_2 < Q_1 P_2 + P_2 Q_2. \tag{3.36}$$

Now consider a shortest path $P_1 Q_2$, an arc of a shortest path γ_2. From the definition of shortest path and (3.36) it follows that

$$\begin{aligned} Q_1 Q_2 &= P_1 P_2 + P_2 Q_2 = l(\overline{P_1 P_2}) + P_2 Q_2 \\ &= P_1 Q_1 + Q_1 P_2 + P_2 Q_2 > P_1 Q_1 + Q_1 Q_2, \end{aligned}$$

which contradicts the triangle inequality.   □

3.5.4 Complete Surfaces

Definition 3.5.2. A surface Φ is *geodesically complete* if each its geodesic segments can be extended limitlessly.

Theorem 3.5.6 (Hopf–Rinow). *If a surface Φ is geodesically complete, then any two of its points can be joined by a shortest path.*

Proof (see [Miln]). Let P and Q be points on Φ, $P \neq Q$, $PQ = a$. Take a circle Γ_δ in a canonical neighborhood V_P of a point P with radius δ. Since the set Γ_δ is compact, there is a point P_δ on Γ_δ for which $P_\delta Q$ is equal to the distance from Q to the set Γ_δ. Draw a geodesic $\gamma(t)$ through the points P and Q, where t is an arc length parameter counting from P. Prove that $\gamma(a) = Q$ and that an arc of $\gamma(t)$ for $0 \leq t \leq a$ is a shortest path joining P and Q. Since any curve with endpoints P and Q intersects Γ_δ, then by definition of the distance and in view of Lemma 3.5.2, the following equality is true:

$$\delta + P_\delta Q = PQ = a. \tag{3.37}$$

Let F be the set of real $t \in (0, a]$ satisfying the equality

$$t + \gamma(t)Q = PQ. \tag{3.38}$$

The set F is nonempty, as is seen from (3.37); it is closed in view of continuity of metrical function ρ_Φ; and it is connected in view of the triangle inequality. Suppose that $t_0 = \sup_{t \in F} t$. If $t_0 = a$, then the theorem is proved.

Thus we suppose that $t_0 < a$, and this will lead to a contradiction. Take a circle Γ_1 of radius δ_1 in a canonical neighborhood $V_{\gamma(t_0)} = V_0$ of the point $\gamma(t_0)$, and select δ_1 such that $\delta_1 < \min(t_0, a - t_0)$ holds. Take a point $P_{\delta_1} = P_1$ on Γ_1 such that

$$\delta_1 + P_1 Q = \gamma(t_0)Q. \qquad (3.39)$$

The existence of the above point P_1 can be proved similarly to how the existence of P_δ was proved. Now two cases are possible:

$$(1) \qquad P_1 \in \gamma, \qquad\qquad (2) \qquad P_1 \notin \gamma.$$

In the first case, from (3.38) and (3.39) it follows that $t_0 + \delta_1 \in F$, contrary to the definition of t_0. So, one needs to study only the second case. In this case, by

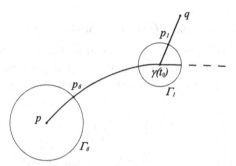

Figure 3.5. Geodesically complete surface.

Lemma 3.5.2,

$$\gamma(t_0 - \delta)P_1 < 2\delta, \qquad (3.40)$$

and from (3.38)–(3.40) follows

$$
\begin{aligned}
PQ &\le QP_1 + P_1\gamma(t_0 - \delta_1) + \gamma(t_0 - \delta_1)P \\
&< QP_1 + 2\delta_1 + t_0 - \delta_1 = t_0 + \delta_1 + \gamma(t_0)Q = PQ,
\end{aligned}
$$

i.e., $PQ < PQ$. This contradiction implies that $t_0 = a$, and the theorem is proved. $\square$

Remark 3.5.2. The assumption of geodesic completeness in Theorem 3.5.6 is necessary. Indeed, if we delete at least one point O from the Euclidean plane $\mathbb{R}^2$, then the points P_1 and P_2, which lie on a same straight line with O, cannot be joined by a shortest path if O is located between P_1 and P_2. In addition to the *geodesic completeness* of a surface one can also define *metrical completeness*, i.e., completeness with respect to the metrical function $\rho_\Phi(P, Q)$. It turns out that both these notions of completeness are equivalent.

Theorem 3.5.7. *A surface* Φ *is geodesically complete if and only if it is metrically complete.*

Proof. Assume that Φ is metrically complete. Then the property of geodesic completeness follows from well-known theorems of the theory of ordinary differential equations. Conversely, let Φ be a geodesically complete surface. Take a bounded and closed set F on Φ in the sense of the metric $\rho_\Phi(P, Q)$. We prove that F is compact. Denote by d the diameter of F. Let $Q \in F$. Construct a closed disk K of radius d in the plane $T\Phi_Q$. Since Φ is geodesically complete, $\exp_Q$ is defined on the whole of K. The set $\exp_Q K$, as the continuous image of a compact set, is also a compact set. In view of Theorem 3.5.6, the set F belongs to $\exp_Q K$; consequently, it is also compact. $\square$

Note that from Theorems 3.5.6 and 3.5.7 it follows that for any point Q on a complete surface Φ the mapping $\exp_Q$ is defined on the whole plane $T\Phi_Q$ and maps it onto all of Φ.

3.5.5 Convex Regions on a Complete Surface

Consider the form taken by the well-known notion of a *convex region* in Euclidean geometry. Let D be some closed region. For simplicity, assume it to be homeomorphic to a disk. Define the *distance between points P and Q of a region D with respect to D* by the rule

$$\rho_D(P, Q) = \inf_{c \in \mathbf{L}_{PQ}(D)} l(c), \tag{3.41}$$

where $\mathbf{L}_{PQ}(D)$ is the *set of all curves in D joining P with Q*. A curve joining P and Q whose length is $\rho_D(P, Q)$ is called a *shortest path in the region D*. A region D on a surface Φ is *geodesically convex* if any shortest path in D is a geodesic on Φ. It turns out that geodesically convex regions, similarly to convex regions in the Euclidean plane $\mathbb{R}^2$, can be characterized locally.

Theorem 3.5.8. *A region D is geodesically convex if and only if the geodesic curvature of its boundary at each point is nonnegative (in the direction of D).*

We leave it to reader as an exercise to find a proof of this theorem.

Definition 3.5.3. A region D on a surface is called *convex* if each shortest path in D is at the same time a shortest path on Φ.

Theorem 3.5.9. *Find an example of a surface and a region on it that is geodesically convex, but not convex.*

Definition 3.5.4. A region D on a surface Φ is called *totally convex* if any shortest path on Φ with ends in D lies entirely in D.

Problem 3.5.2. Find an example of a surface and a region on it that is convex but not totally convex.

Finally, we shall present the notion of an absolutely convex region on a surface.

Definition 3.5.5. A region D is called *absolutely convex* if any arc of a geodesic γ with ends in D lies entirely in D.

Problem 3.5.3. Prove that on a compact surface no absolutely convex region exists.

Definition 3.5.6. A half-geodesic with initial point at P is called a *ray* with vertex P if any of its arcs is a shortest path.

Problem 3.5.4. Prove that on any complete open surface Φ of class C^2 through any point $P \in \Phi$ there passes at least one ray.

Definition 3.5.7. Let r_1 be some ray on a complete open surface Φ of class C^2. We say that a ray r_2 with the vertex P is a *co-ray* for r_1 if it is the limit of the shortest paths $P P_n$ as $P_n \in r_1$ tends to infinity.

Problem 3.5.5. Prove that through every point $P \in \Phi$ there exists a co-ray for a given ray r_1.

Problem 3.5.6. Let r_2 be the co-ray for some ray r_1, and Q the inner point on the ray r_2. Prove that through the point Q there is a unique co-ray for the ray r_1.

Example 3.5.1. Consider the paraboloid of revolution Φ: $z = x^2 + y^2$ in $\mathbb{R}^3$. Prove that *each region $D(a) \subset \Phi$, defined by the inequality $0 \leq z \leq a$, is an absolutely convex region.* A paraboloid Φ divides $\mathbb{R}^3$ into two regions, one of which, namely the one containing the positive axis OZ, is a convex region. Denote this region by B. Let $\vec{n}(P)$ be a normal to Φ at the point P. Assume that it is directed inside of B. Then $\vec{n}(P)$ forms an acute angle with the positive axis OZ.

Take an arc of an arbitrary geodesic, $\gamma(t)$ ($a \leq t \leq b$), on Φ. We study the behavior of a function z on the arc $\gamma(t)$. It turns out that a maximum of z on γ is reached only at the endpoints (or at one of the endpoints). Assume the opposite. Let the function $z(t)$ take its maximum at a point $\gamma(t_0)$. Pass the plane $z = z(t_0) = z_0$ through $\gamma(t_0)$. Then the curve $\gamma(t)$ in some neighborhood of $\gamma(t_0)$ does not lie entirely above the plane $z = z_0$. Thus the main normal $\vec{v}(t_0)$ to γ at the point $\gamma(t_0)$ forms an angle not smaller than $\pi/2$ with the positive semiaxis OZ, and the normal $\vec{n}(\gamma(t_0))$ forms an acute angle with the same semiaxis. Hence, the normal $\vec{n}(\gamma(t_0))$ to Φ and the main normal $\vec{v}(t_0)$ to γ at $\gamma(t_0)$ are not parallel, which contradicts the definition of a geodesic.

Now we can prove the absolutely convexity of the region $D(a)$ for any $a > 0$. Let $P \in D(a)$ and $Q \in D(a)$. Denote by $\gamma(t)$ an arc of an arbitrary geodesic with endpoints P and Q. Assume that $\gamma(t)$ does not lie entirely in $D(a)$. Then, since z is not greater than a at P and Q by definition of $D(a)$, the function z takes its maximal value at some inner point of the arc of the geodesic $\gamma(t)$, which contradicts the above statement. Hence, $D(a)$ is an absolutely convex region for any $a > 0$.

3.6 Special Coordinate Systems

3.6.1 Riemannian Normal Coordinate System

Let P be an arbitrary point on the surface Φ. Consider the rectangular coordinate system u^1, u^2 in the plane $T\Phi_P$. In view of Lemma 3.4.3, there is δ such that the disk $B(P, \delta)$ with center at P and radius δ is uniquely mapped onto some neighborhood W of P on Φ under the exponential map. Let $Q_1 \in B(P, \delta)$ and $Q = \exp_P(\overrightarrow{PQ_1})$. Introduce a coordinate system u^1, u^2 in W, letting the coordinates of the point Q be equal to the coordinates of the point P. Then P has zero coordinates, and $\vec{r}_1(0, 0)$ and $\vec{r}_2(0, 0)$ are unit and mutually orthogonal vectors. Thus

$$g_{11}(0, 0) = g_{22}(0, 0) = 1, \qquad g_{12}(0, 0) = 0.$$

The equations of geodesics on Φ passing through P will be written in the form $u^1 = \alpha^1 t, u^2 = \alpha^2 t$, where t is proportional to an arc length parameter and is equal to it when $(\alpha^1)^2 + (\alpha^1)^2 = 1$. Substituting these functions in the equations of geodesics, we obtain

$$\Gamma^i_{jk}(\alpha^1 t, \alpha^2 t)\alpha^j\alpha^k = 0 \qquad (i = 1, 2). \tag{3.42}$$

Multiplying (3.42) by g_{ip} and summing over i, we obtain

$$\Gamma_{jk,p}(\alpha^1 t, \alpha^2 t)\alpha^j\alpha^k = 0 \qquad (p = 1, 2). \tag{3.43}$$

Let $t = 0$. Then

$$\Gamma_{jk,p}(0, 0)\alpha^j\alpha^k = 0 \qquad (p = 1, 2). \tag{3.44}$$

From (3.44), in view of the arbitrariness of the real α^1 and α^2, we have

$$\Gamma_{ij,k} = 0 \qquad (i, j, k = 1, 2). \tag{3.45}$$

For calculating the second derivatives of the metric tensor components, differentiate (3.43) with respect to t and assume that $t = 0$. Then

$$\frac{\partial \Gamma_{jk,p}}{\partial u^i}\alpha^i\alpha^j\alpha^k = 0 \qquad (p = 1, 2). \tag{3.46}$$

Write down (3.46) in detail:

$$(\alpha^1)^3\frac{\partial \Gamma_{11,p}}{\partial u^1} + (\alpha^1)^2\alpha^2\left(\frac{\partial \Gamma_{11,p}}{\partial u^2} + \frac{\partial \Gamma_{12,p}}{\partial u^1}\right) + \alpha^1(\alpha^2)^2\left(\frac{\partial \Gamma_{22,p}}{\partial u^1} + \frac{\partial \Gamma_{12,p}}{\partial u^2}\right)$$
$$+ (\alpha^2)^3\frac{\partial \Gamma_{22,p}}{\partial u^2} = 0 \qquad (p = 1, 2). \tag{3.47}$$

From (3.47) we obtain **eight** equations:

$$\frac{\partial \Gamma_{11,p}}{\partial u^1}(0,0) = \frac{\partial^2 g_{1p}}{(\partial u^1)^2} - \frac{1}{2}\frac{\partial^2 g_{11}}{\partial u^1 \partial u^p} = 0 \qquad (p=1,2), \qquad (3.48)$$

$$\frac{\partial \Gamma_{11,p}}{\partial u^2}(0,0) + 2\frac{\partial \Gamma_{12,p}}{\partial u^1}(0,0) = \frac{\partial^2 g_{1p}}{\partial u^1 \partial u^2} - \frac{1}{2}\frac{\partial^2 g_{11}}{\partial u^2 \partial u^p}$$

$$+ \frac{\partial^2 g_{1p}}{\partial u^1 \partial u^2} + \frac{\partial^2 g_{2p}}{(\partial u^1)^2} - \frac{\partial^2 g_{12}}{\partial u^1 \partial u^p} = 0 \qquad (p=1,2), \qquad (3.49)$$

$$\frac{\partial \Gamma_{22,p}}{\partial u^1}(0,0) + 2\frac{\partial \Gamma_{12,p}}{\partial u^2}(0,0) = \frac{\partial^2 g_{2p}}{\partial u^1 \partial u^2} - \frac{1}{2}\frac{\partial^2 g_{22}}{\partial u^1 \partial u^p}$$

$$+ \frac{\partial^2 g_{1p}}{(\partial u^2)^2} + \frac{\partial^2 g_{2p}}{\partial u^1 \partial u^2} - \frac{\partial^2 g_{12}}{\partial u^2 \partial u^p} = 0 \qquad (p=1,2), \qquad (3.50)$$

$$\frac{\partial \Gamma_{22,p}}{\partial u^2}(0,0) = \frac{\partial^2 g_{2p}}{(\partial u^2)^2} - \frac{1}{2}\frac{\partial^2 g_{22}}{\partial u^2 \partial u^p} = 0 \qquad (p=1,2). \qquad (3.51)$$

We add to these equations the expression for the Gaussian curvature K of a surface Φ at P. In view of (3.45), we obtain

$$K = \frac{1}{2}\left(2\frac{\partial^2 g_{12}}{\partial u^1 \partial u^2} - \frac{\partial^2 g_{11}}{(\partial u^2)^2} - \frac{\partial^2 g_{22}}{(\partial u^1)^2}\right). \qquad (3.52)$$

Solving the system (3.48)–(3.52), we obtain

$$\frac{\partial^2 g_{11}}{(\partial u^1)^2}(0,0) = \frac{\partial^2 g_{11}}{\partial u^1 \partial u^2}(0,0) = \frac{\partial^2 g_{12}}{(\partial u^1)^2}(0,0)$$

$$= \frac{\partial^2 g_{22}}{\partial u^1 \partial u^2}(0,0) = \frac{\partial^2 g_{12}}{(\partial u^2)^2}(0,0) = 0,$$

$$\frac{\partial^2 g_{11}}{(\partial u^2)^2}(0,0) = \frac{\partial^2 g_{22}}{(\partial u^1)^2}(0,0) = -\frac{2}{3}K(0,0),$$

$$\frac{\partial^2 g_{12}}{\partial u^1 \partial u^2}(0,0) = \frac{1}{3}K(0,0). \qquad (3.53)$$

From (3.53) it follows that

$$g_{11} = 1 - \frac{1}{3}K(0,0)(u^2)^2 + \bar{o}((u^1)^2 + (u^2)^2),$$

$$g_{12} = \frac{1}{3}K(0,0)u^1 u^2 + \bar{o}((u^1)^2 + (u^2)^2), \qquad (3.54)$$

$$g_{22} = 1 - \frac{1}{3}K(0,0)(u^1)^2 + \bar{o}((u^1)^2 + (u^2)^2).$$

The formulas (3.54) give us the possibility to compare the length of an arbitrary curve γ_1 on a tangent plane $T\Phi_P$ with the length of its image γ under the exponential map.

3.6.2 Comparison Theorem for Metrics

Let a curve γ_1 lie inside a disk $B(P, \delta)$ in a plane $T\Phi_P$, and let γ be its image under the exponential map. Denote by l_1 the length of γ_1, and by l the length of γ.

Theorem 3.6.1. *There exists δ_1 such that for all $\delta < \delta_0$ the inequality $|l - l_1| \leq Cl_1\delta^2$ holds, where the constant C does not depend on the value of the Gaussian curvature at a point P.*

Proof. Let the equations of the curve γ_1 in Cartesian rectangular coordinates on $T\Phi_P$ be presented by the functions $f_1(t)$ and $f_2(t)$; $u^1 = f_1(t)$, $u^2 = f_2(t)$. Then the equations of the curve γ in *Riemannian normal coordinates* are given by the same functions. Let t be the arc length of the curve γ_1 counting from its initial point. Then

$$[f_1'(t)]^2 + [f_2'(t)]^2 = 1. \tag{3.55}$$

Calculate the length l of the curve γ:

$$l = \int_0^{l_1} \sqrt{(f_1')^2 + (f_2')^2 + \tfrac{1}{3}K(0,0)[2f_1 f_2 f_1' f_2' - f_1^2(f_1')^2 - f_2^2(f_2')^2] + \bar{o}(\delta^2)}\, dt. \tag{3.56}$$

Set

$$A = \frac{1}{3}K(0,0)(2f_1 f_2 f_1' f_2' - f_1^2(f_1')^2 - f_2^2(f_2')^2) + \bar{o}(\delta^2).$$

Then

$$|l - l_1| = \int_0^{l_1} (\sqrt{1 + A} - 1)\, dt = \int_0^{l_1} \frac{A\, dt}{\sqrt{1 + A} + 1}.$$

We estimate $|A|$ from above:

$$|A| \leq \frac{1}{3}|K(0,0)| \cdot 2[f_1^2(f_1')^2 + f_2^2(f_2')^2] + \bar{o}(\delta^2).$$

Since $|f_1'| \leq 1$ and $|f_2'| \leq 1$, for a sufficiently small δ we obtain

$$|A| \leq \frac{1}{3}|K(0,0)|2\delta^2 + \bar{o}(\delta^2) \leq |K(0,0)|\delta^2.$$

We estimate $\sqrt{1 + A} + 1$ from below. In view of the previous inequality, we have $\sqrt{1 + A} + 1 > \frac{3}{2}$. We thus obtain

$$|l - l_1| \leq \frac{2}{3}|K(0,0)|\delta^2 \int_0^{l_1} dt = \frac{2}{3}|K(0,0)|l_1\delta^2. \qquad \Box$$

In particular, let γ_1 be the circle of radius δ with center at P in the plane $T\Phi_P$, and γ the circle of radius δ with center at P on Φ. Then from Theorem 3.6.1 follows $l = 2\pi\delta + \bar{o}(\delta^2)$, and if γ is an arc of a circle with a central angle α, then

$$l = \alpha\delta + \bar{o}(\delta^2). \tag{3.57}$$

Corollary 3.6.1. *Let $A \in B(P, \delta)$, $B \in B(P, \delta)$, $\bar{A} = \exp_P^{-1}(A)$, and $\bar{B} = \exp_P^{-1}(B)$. Then $AB = \bar{A}\bar{B} + \bar{o}(\delta)$.*

Proof. Let $\gamma = \exp_P(\bar{A}\bar{B})$, and $\bar{\gamma} = \exp_P^{-1}(AB)$. Then, in view of Theorem 3.6.1, we have $l = l(\gamma) = \bar{A}\bar{B} + \bar{o}(\delta)$, $\bar{l} = l(\bar{\gamma}) = AB + \bar{o}(\delta)$. But since $AB \leq l$, $\bar{A}\bar{B} < \bar{l}$, then $AB \leq \bar{A}\bar{B} + \bar{o}(\delta)$, $\bar{A}\bar{B} \leq AB + \bar{o}(\delta)$. From the two last inequalities it follows that $AB = \bar{A}\bar{B} + \bar{o}(\delta)$. $\qquad \Box$

3.6.3 Geodesic Polar Coordinates on a Surface

Let P be an arbitrary point on a surface Φ. Introduce a polar coordinate system with the center at a point P in the plane $T\Phi_P$. In view of Lemma 3.4.3, there is $\delta_0 > 0$ such that for all δ $(0 < \delta < \delta_0)$ the exponential map $\exp_P$ is a diffeomorphism of a disk $\tilde{B}(P, \delta) \subset T\Phi_P$ onto some neighborhood $W \subset \Phi$. Introduce coordinates (ρ, φ) in W, assuming coordinates of a point Q equal to the coordinates of its inverse image under the exponential map $\exp_P$. This coordinate system is called a *geodesic polar coordinate system with the origin at P*, and a neighborhood W itself is called a *disk*, which we denote by the same symbol $B(P, \delta)$. The point P is a singular point of this coordinate system: $|\vec{r}_\rho(0, 0)| = 1$, and $|\vec{r}_\varphi(0, 0)| = 0$. Therefore,

$$g_{11}(0, 0) = \langle \vec{r}_\rho, \vec{r}_\rho \rangle = 1, \quad g_{22}(0, 0) = \langle \vec{r}_\varphi, \vec{r}_\varphi \rangle = 0,$$
$$g_{12}(0, 0) = \langle \vec{r}_\rho, \vec{r}_\varphi \rangle = 0. \tag{3.58}$$

In view of Lemma 3.5.1 about the derivative of the length of a family of geodesics, we see that the coordinate curves $\rho = \text{const}$ and $\varphi = \text{const}$ are orthogonal. Thus

$$g_{12}(\rho, \varphi) = 0. \tag{3.59}$$

Set $f = \sqrt{g_{22}(\rho, \varphi)}$. Then from (3.58) it follows that

$$f(0, \varphi) = 0. \tag{3.60}$$

We prove that $f_\rho'(0, \varphi) = 1$. Let $l(\rho, \alpha)$ be an arc length parameter of the circle of radius ρ and center at P and with central angle α on Φ. Then

$$l(\rho, \alpha) = \int_0^\alpha \sqrt{g_{22}(\rho, \varphi)} \, d\varphi.$$

Hence from (3.57) follows the identity

$$\int_0^\alpha \sqrt{g_{22}(\rho, \varphi)} \, d\varphi = \alpha\rho + \bar{o}(\rho).$$

Differentiating the last equality with respect to ρ and α, assuming then $\rho = 0$, we get

$$\frac{d}{d\rho}\left(g_{22}(\rho, \varphi)\right)\big|_{\rho=0} = 1,$$

or

$$f_\rho'(0, \varphi) = 1. \tag{3.61}$$

We now calculate the Gaussian curvature K of the surface Φ. In view of equalities (3.58) and (3.59), we obtain

$$K(\rho, \varphi) = -f_{\rho\rho}''(\rho, \varphi)/f(\rho, \varphi). \tag{3.62}$$

We rewrite (3.62) in the following form:

$$f''_{\rho\rho}(\rho, \varphi) + K(\rho, \varphi) f(\rho, \varphi) = 0, \tag{3.63a}$$

$$f(0, \varphi) = 0, \quad f'_\rho(0, \varphi) = 1. \tag{3.63b}$$

Equation (3.63a) and initial conditions (3.63b) show us that if $K(\rho, \varphi)$ is a known function of coordinates (ρ, φ), then $g_{22}(\rho, \varphi)$ is uniquely defined and all coefficients of the metric tensor can be obtained. For instance, let $K(\rho, \varphi) = \pm a^2$. Then $g_{11} = 1$, $g_{12} = 0$, $g_{22} = \frac{1}{a^2}\sin^2(a\rho)$ or $g_{22} = \frac{1}{a^2}\sinh^2(a\rho)$ in the case of $K = -a^2$.

3.6.4 Semigeodesic Coordinate System on a Surface

Take on a surface Φ a closed arc of a geodesic γ without points of self-intersection. Introduce on γ an arc length parameterization $\gamma(t)$ ($0 \le t \le b$), counting from some point. Pass a geodesic $\sigma(t, s)$ through every point $\gamma(t)$ orthogonal to γ, where s is an arc length parameter on $\sigma(t, s)$, starting from the point $\gamma(t)$. Prove that there is $\delta_0 > 0$ such that for $|s| < \delta_0$ none of the geodesics $\sigma(t, s)$ intersect for a sufficiently small extension. First, prove that two sufficiently nearby geodesics $\sigma(t, s)$ do not intersect on a sufficiently small extension.

Take some point $\gamma(t)$, and let any coordinates u^1, u^2 be defined in its neighborhood. Let $\gamma(t)$ have the equations $u^i = u^i(t)$ ($i = 1, 2$), and let $\sigma(t, s)$ have the equations $u^i = h^i(t, s)$. Then

$$h^i(t, s) = u^i(t) \qquad (i = 1, 2). \tag{3.64}$$

Prove that the determinant $\Delta = \begin{vmatrix} \frac{\partial h^1}{\partial t} & \frac{\partial h^2}{\partial t} \\ \frac{\partial h^1}{\partial s} & \frac{\partial h^2}{\partial s} \end{vmatrix}$ is nonzero for $s = 0$. Set $\lambda^i = \frac{\partial h^i}{\partial t}\big|_{s=0}$ and $\mu^i = \frac{\partial h^i}{\partial s}\big|_{s=0}$. Then λ^i are coordinates of a unit vector $\vec{\lambda}$ that is tangent to $\gamma(t)$ at the point $\gamma(t)$, and μ^i are coordinates of a unit vector $\vec{\mu}$ that is tangent to the geodesic $\sigma(t, s)$ at the point $\sigma(t, 0) = \gamma(t)$. Since $\vec{\lambda}$ and $\vec{\mu}$ are mutually orthogonal vectors by their construction, then $\Delta \ne 0$. From this and the implicit function theorem it follows that for each t_1 there exist real numbers $\delta(t_1)$ and $\varepsilon(t_1)$ such that for $t_1 - \varepsilon(t_1) < t < t_1 + \varepsilon(t_1)$ the arcs of the geodesics $\sigma(t, s)$ do not intersect. Moreover, all points $\sigma(t, s)$ for $t_1 - \varepsilon(t_1) < t < t_1 + \varepsilon(t_1)$, $|s| < \delta(t_1)$ form a region on Φ. Since the closed arc γ is compact, there exist real numbers δ_1 and ε_1 such that the arcs of geodesics $\sigma(t_1, s)$ and $\sigma(t_2, s)$ do not intersect when $|s| < \delta_1$ and $|t_2 - t_1| < \varepsilon_1$.

Now let $\delta_2 = \min\{\rho(\gamma(t), \gamma(t + t')): t, t' \in [a, b], t' \ge \varepsilon'\}$. Define $\delta_0 = \frac{1}{2}\min\{\delta_1, \delta_2\}$. It is not difficult to check that δ_0 is the required real number. Consider the region on the surface Φ consisting of the points $\sigma(t, s)$ for $a \le t \le b$. Introduce coordinates in this region, assuming $u^1 = s$, $u^2 = t$. The obtained coordinates are called *semigeodesic coordinates*.

Remark 3.6.1. In the construction of semigeodesic coordinates one can, generally speaking, take an arbitrary regular curve instead of a geodesic γ.

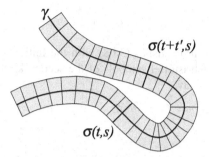

Figure 3.6. Semigeodesic coordinates on a surface.

We now study the structure of the coefficients of the first fundamental form of a surface in a semigeodesic coordinate system. The coordinate curves $u^1 = $ const and $u^2 = $ const intersect in a right angle. This statement follows from the construction of the coordinate system and from Lemma 3.5.1 about the derivative of the length of a family of geodesics. Therefore,

$$g_{21}(u^1, u^2) = 0. \tag{3.65}$$

Furthermore, since u^1 is the length of the curve $u^2 = $ const, then

$$g_{11}(u^1, u^2) = 1. \tag{3.66}$$

Thus,

$$ds^2 = (du^1)^2 + g_{22}(u^1, u^2)(du^2)^2. \tag{3.67}$$

Assuming $f(u^1, u^2) = \sqrt{g_{22}}$, we obtain, as in Section 3.6.3

$$\frac{\partial^2 f}{(\partial u^1)^2} + K(u^1, u^2) f(u^1, u^2) = 0. \tag{3.68}$$

However, the initial conditions are different. Since for $u^1 = 0$, u^2 is the length of γ, then

$$\sqrt{g_{22}(0, u^2)} = f(0, u^2) = 1. \tag{3.69}$$

Finally, $u^1 = 0$, $u^2 = t$ is a geodesic; consequently,

$$\Gamma_{22}^1(0, u^2) = \Gamma_{22}^2(0, u^2) = 0. \tag{3.70}$$

Using the expressions of Christoffel symbols of the first kind, we obtain

$$-f(0, u^2)\frac{\partial f}{\partial u^1}(0, u^2) = 0,$$

and from this follows

$$\frac{\partial f}{\partial u^1}(0, u^2) = 0. \tag{3.71}$$

So the function $f(u^1, u^2)$ satisfies (3.68) with the initial conditions (3.69) and (3.71). We calculate Christoffel symbols of the second kind for a semigeodesic coordinate system. We have

$$g_{22} = G(u^1, u^2) = f^2(u^1, u^2), \qquad g_{12} = g_{21} = 0, \qquad g_{11} = 1.$$

Thus

$$\Gamma^1_{11} = \Gamma^2_{11} = \Gamma^1_{12} = \Gamma^1_{21} = 0, \qquad \Gamma^1_{22} = \Gamma_{22,1} = -\frac{1}{2}G_{u^1} = -ff_{u^1},$$

$$\Gamma^2_{21} = \Gamma^2_{12} = \frac{2}{f^2}ff_{u^1} = \frac{f_{u^1}}{f}, \qquad \Gamma^2_{22} = \frac{1}{f^2}\Gamma_{22,2} = \frac{f_{u^2}}{f}. \tag{3.72}$$

Furthermore, we obtain a formula for the geodesic curvature k_g of an arbitrary curve c. Let $u^1 = u^1(t)$, $u^2 = u^2(t)$ be the equations of c, and t an arc length parameter on c, counting from some point. Then we have the equalities

$$|\dot{c}(t)| = 1, \qquad \left(\dot{c}(t), \frac{D}{dt}\dot{c}(t) \right) = 0. \tag{3.73}$$

If we denote by μ^1 and μ^2 the coordinates of the vector $\frac{D}{dt}\dot{c}(t)$ in a local basis, then the system (3.73) in semigeodesic coordinates is written in the following form:

$$\left(\frac{du^1}{dt} \right)^2 + G\left(\frac{du^2}{dt} \right)^2 = 1, \tag{3.74}$$

$$\mu^1 \frac{du^1}{dt} + G\mu^2 \frac{du^2}{dt} = 0. \tag{3.75}$$

Lemma 3.6.1. *The geodesic curvature k_g of a curve c is expressed by the formula*

$$k_g = \sqrt{G} \left| \mu^1 \frac{du^2}{dt} - \mu^2 \frac{du^1}{dt} \right|. \tag{3.76}$$

Proof. Raise the right-hand side to the second power. In view of (3.74), (3.75), and (3.21) we obtain

$$G\left[\left(\mu^1 \frac{du^2}{dt} \right)^2 + \left(\mu^2 \frac{du^1}{dt} \right)^2 - 2\mu^1\mu^2 \frac{du^1}{dt}\frac{du^2}{dt} \right]$$

$$= G\left[\left(\mu^1 \frac{du^2}{dt} \right)^2 + \left(\mu^2 \frac{du^1}{dt} \right)^2 \right.$$

$$\left. - \frac{\mu^1}{G}\frac{du^1}{dt}\left(-\mu^1 \frac{du^1}{dt} \right) - \mu^2 \frac{du^2}{dt}\left(-G\mu^2 \frac{du^2}{dt} \right) \right]$$

$$= G\left[\frac{(\mu^1)^2}{G}\left(G\left(\frac{du^2}{dt} \right)^2 + \left(\frac{du^1}{dt} \right)^2 \right) + (\mu^2)^2 \left(\left(\frac{du^1}{dt} \right)^2 + G\left(\frac{du^2}{dt} \right)^2 \right) \right]$$

$$= (\mu^1)^2 + G(\mu^2)^2 = \left| \frac{D}{dt}\dot{c}(t) \right|^2 = k_g^2. \qquad \square$$

Now substitute the expressions for u^1 and u^2, see (3.8), in (3.76). Then the geodesic curvature is

$$k_g = \left| f \left[\frac{du^2}{dt} \frac{d^2 u^1}{dt^2} - f f_{u^1} \left(\frac{du^2}{dt} \right)^2 \right] \right.$$
$$\left. - f \frac{du^1}{dt} \left[\frac{d^2 u^2}{dt^2} + 2 \frac{f_{u^1}}{f} \frac{du^1}{dt} \frac{du^2}{dt} + \frac{f_{u^2}}{f} \left(\frac{du^2}{dt} \right)^2 \right] \right|.$$

Simplifying the sums of the above terms yields

$$- f^2 f_{u^1} \left(\frac{du^2}{dt} \right)^3 - f_{u^1} \left(\frac{du^1}{dt} \right)^2 \frac{du^2}{dt}$$
$$= - \frac{du^2}{dt} f_{u^1} \left[f^2 \left(\frac{du^2}{dt} \right)^2 + \left(\frac{du^1}{dt} \right)^2 \right] = - f_{u^1} \frac{du^2}{dt}.$$

Consequently,

$$k_g = \left| f \left(\frac{d^2 u^1}{dt^2} \frac{du^2}{dt} - \frac{d^2 u^2}{dt^2} \frac{du^1}{dt} \right) \right.$$
$$\left. - f_{u^1} \left(\left(\frac{du^1}{dt} \right)^2 + 1 \right) \frac{du^2}{dt} - f_{u^2} \left(\frac{du^2}{dt} \right)^2 \frac{du^1}{dt} \right|. \qquad (3.77)$$

3.7 Gauss–Bonnet Theorem and Comparison Theorem for the Angles of a Triangle

Consider a region W on a complete surface Φ that is homeomorphic to a disk and that lies in some coordinate neighborhood U with coordinates (u^1, u^2). Introduce an *orientation in W, induced by coordinates* (u^1, u^2); i.e., define a basis $\vec{r}_1 = \vec{r}_{u^1}$, $\vec{r}_2 = \vec{r}_{u^2}$ at each point of W. Assume that the boundary of W is a curve c of class C^2. Take a field of normals $\vec{a}$ along c, directed inside of W. Introduce a parameterization on c such that the ordered pair of vectors $\{\vec{a}, \dot{c}(t)\}$ forms a basis equivalent to the basis $\{\vec{r}_1, \vec{r}_2\}$. Define the sign of the geodesic curvature k_g of the curve $c(t)$ in the following way.

If the basis $\{ \frac{D}{dt} \dot{c}(t), \dot{c}(t) \}$ is equivalent to the basis $\{\vec{a}, \dot{c}(t)\}$, then suppose $k_g > 0$, and $k_g < 0$ for the opposite case. If μ^1, μ^2 are the components of $\frac{D}{dt} \dot{c}(t)$ in this basis $\{\vec{r}_1, \vec{r}_2\}$, and (c^1, c^2) are the components of $\dot{c}(t)$ in the same basis, then the signs of the curvature k_g and the determinant $\begin{vmatrix} \mu^1 & \mu^2 \\ c^1 & c^2 \end{vmatrix} = \mu^1 c^2 - \mu^2 c^1$ coincide. So,

$$\operatorname{sign} k_g = \operatorname{sign}(\mu^1 c^2 - \mu^2 c^1). \qquad (3.78)$$

From the geometrical point of view, the definition of $\operatorname{sign} k_g$ is equivalent to the condition that *if k_g is positive at each point on $c(t)$, then W is a geodesically convex region.*

We move now to the formulation of the Gauss–Bonnet theorem. Let the boundary of a region W be a piecewise smooth curve with a finite number of angular points $A_1, \ldots, A_n$ joined in consecutive order by regular curves $c_1, \ldots, c_n$ of class C^2. Denote by α_i an *interior angle from the side of the region* W at the vertex A_i $(i = 1, \ldots, n)$. Then the following theorem holds.

Theorem 3.7.1. (Gauss–Bonnet.) *The following equality is satisfied:*

$$\sum_{i=1}^{n} \int_{c_i} k_g \, dt + \sum_{i=1}^{n} (\pi - \alpha_i) + \iint_{W} K \, dS = 2\pi, \qquad (3.79)$$

where t is an arc length parameter, dS an element of the area on Φ, and K the Gaussian curvature.

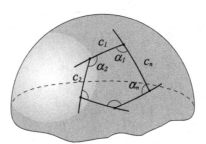

Figure 3.7. Gauss–Bonnet theorem.

Proof. First consider the case that the boundary ∂W of W does not contain angular points and it is possible to introduce semigeodesic coordinates in the whole region W. The magnitude $k_g(t)$, in view of the definition of the sign of k_g in (3.78), the equality $\sqrt{G} = f$, and (3.77), can be written in the following form:

$$k_g \, dt = -d\left(\arctan \sqrt{G} \frac{(u^2)'}{(u^1)'} \right) - (u^2)'(\sqrt{G})_{u^1} \, dt.$$

Thus

$$\int_c k_g \, dt = - \int_c d\left(\arctan \sqrt{G} \frac{(u^2)'}{(u^1)'} \right) - \int_c (\sqrt{G})_{u^1} \, du^2.$$

Since arctan is a multivalued function, its values corresponding to the same value of an argument differ by a multiple of π. Hence

$$- \int_c d\left(\arctan \sqrt{G} \frac{(u^2)'}{(u^1)'} \right) = \pi m, \qquad (3.80)$$

where m is some integer. The second term $\int_c (\sqrt{G})_{u^1} \, du^2$, by Green's formula and (3.63a), is transformed to the following form:

$$\int_c (\sqrt{G})_{u^1} \, du^2 = \iint_W (\sqrt{G})_{u^1 u^1} \, du^1 \, du^2$$

$$= \iint_W \frac{(\sqrt{G})_{u^1 u^1}}{\sqrt{G}} \sqrt{G} \, du^1 \, du^2 = - \iint_W K \, dS.$$

So we obtain

$$\int_c k_g \, dt = m\pi - \iint_W K \, dS.$$

It remains now only to derive the value of m. If the function $f(u^1 u^2)$ were identically 1, then the magnitude $m\pi$ would be equal to the angle of rotation of the tangent to the curve $c(t)$ vector while moving around this curve. Obviously, this angle is 2π, i.e., the value of m is then equal to 2. Consider the integral

$$\int_\gamma -d\left(\arctan \frac{a(u^1 u^2)(u^2)'}{(u^1)'} \right).$$

This integral continuously depends on $a(u^1 u^2)$ and is equal to 2π when $a(u^1 u^2) = 1$. Consequently, this integral is 2π for any function $a(u^1 u^2)$ satisfying the positivity condition, in particular, for $a = f$. We thus obtain that $\int_c k_g \, dt + \iint_W K \, dS = 2\pi$. The Gauss–Bonnet theorem for this case is proved.

Now consider the case that $\partial W = c(t)$ contains angular points. At each angular point A_i the tangent vector $c(t)$ turns through the angle $(\pi - \alpha_i)$ (see Section 3.4). Hence in this case, instead of the integral $\int_c k_g \, dt$, we must write

$$\sum_{i=1}^n \int_{c_i} k_g \, dt + \sum_{i=1}^n (\pi - \alpha_i),$$

and again we obtain the Gauss–Bonnet formula (3.79).

Finally, we need not assume the existence of a global semigeodesic coordinate system on W: Consider the particular case that W can be divided into two regions W_1 and W_2 for which the Gauss–Bonnet formula holds. Assume for simplicity that the boundary of W is a regular curve $c(t)$ of class C^2, but a curve γ_1 dividing W onto W_1 and W_2, is also a regular curve of class C^2. Denote by A and B the endpoints of the curve γ_1, and by α_1, α_2 and β_1, β_2 the interior angles of the regions W_1 and W_2 at the points A and B, respectively, and by c_1 and c_2 the arcs of the curve c. Then by our assumption, we have

$$\int_{c_1} k_g \, dt + \int_{\gamma_1} k_g^1 \, dt + \pi - \alpha_1 + \pi - \alpha_2 + \iint_{W_1} K \, dS = 2\pi,$$

$$\int_{c_2} k_g \, dt + \int_{\gamma_1} k_g^2 \, dt + \pi - \beta_1 + \pi - \beta_2 + \iint_{W_2} K \, dS = 2\pi.$$

Here k_g^1 is the geodesic curvature of the curve γ_1 whose sign is determined by W_1, and k_g^2 is the geodesic curvature of the same curve γ_1 whose sign is defined by

W_2. Obviously, $k_g^1 + k_g^2 = 0$. Taking the sum of left and right terms of the two last two equalities, we obtain

$$\int_{c_1} k_g \, dt + \int_{c_2} k_g \, dt + 4\pi - (\alpha_1 + \alpha_2 + \beta_1 + \beta_2) + \iint_{W_1} K \, dS + \iint_{W_2} K \, dS = 4\pi.$$

Since $\alpha_1 + \alpha_2 + \beta_1 + \beta_2 = 2\pi$, we have $\int_c k_g \, dt + \iint_W K \, dS = 2\pi$.   □

In the general case, the proof of the Gauss–Bonnet theorem can be obtained by induction on the number of regions dividing W.

We now deduce some corollaries of the Gauss–Bonnet theorem.

Corollary 3.7.1. Take on a surface Φ a triangle $\triangle$, composed of geodesics, and assume that a region D bounded by $\triangle$ is homeomorphic to a disk. Denote by α_1, α_2, and α_3 the interior angles of the triangle. Apply the Gauss–Bonnet formula to D:

$$\iint_D K \, dS + (\pi - \alpha_1) + (\pi - \alpha_2) + (\pi - \alpha_3) = 2\pi,$$

or

$$\alpha_1 + \alpha_2 + \alpha_3 = \pi + \iint_D K \, dS. \tag{3.81}$$

If $K \equiv 0$, then we obtain a well-known theorem from elementary geometry. For $K > 0$ we see that the sum of angles of the triangle is greater than π, and for $K < 0$ the sum is smaller than π.

Corollary 3.7.2. If a surface Φ is homeomorphic to a sphere, then its integral curvature is 4π.

Proof. Transfer a closed smooth curve γ onto Φ, dividing it into two regions D_1 and D_2, each of them is homeomorphic to a disk. Apply the Gauss–Bonnet formula to D_1 and D_2:

$$\iint_{D_1} K \, dS + \int_\gamma k_g^1 \, dt = 2\pi, \qquad \iint_{D_2} K \, dS + \int_\gamma k_g^2 \, dt = 2\pi. \tag{3.82}$$

Here we denote by k_g^1 the geodesic curvature of the curve γ whose sign is determined with respect to the region D_1, and by k_g^2 the geodesic curvature of the same curve γ, but with the sign of k_g^2 determined with respect to D_2. Therefore,

$$k_g^1 + k_g^2 = 0. \tag{3.83}$$

Summing the two formulas of (3.82), in view of (3.83), we obtain

$$\iint_{D_1} K \, dS + \iint_{D_2} K \, dS = \iint_\Phi K \, dS = 4\pi.$$ □

Problem 3.7.1. Prove that the integral curvature of any surface homeomorphic to a torus is 0.

Remark 3.7.1. The total Gaussian curvature of a compact surface Φ is

$$\iint_\Phi K \, dS = 2\pi \chi(\Phi),$$

where $\chi(\Phi)$ is the Euler characteristic (see Remark 2.7.1).

Problem 3.7.2 (Sine theorem for small triangles). Let $\triangle ABC$ be a triangle on the regular surface of class C^2, and $AC = \delta$. Then

$$BC = \sin\alpha \cdot \delta/\sin(\alpha+\gamma) + \bar{o}(\delta), \qquad AB = \sin\gamma \cdot \delta/\sin(\alpha+\gamma) + \bar{o}(\delta),$$

where α and γ are the angles of $\triangle ABC$ at the vertices A and C, respectively.

Solution. Introduce on AC the arc length parameterization $c(s)$, counting from the point A, $0 \le s \le \delta$. Bring a geodesic $\sigma(s)$ through the point $c(s)$ forming an angle γ with AC, and denote by $B(s)$ the point of intersection of $\sigma(s)$ with AB. Let $t(s) = AB(s)$, $l(s) = A(s)B(s)$ and denote by $\beta(s)$ the angle of $\triangle ABC$ at the vertex $B(s)$. Then, in view of Lemma 3.5.1, we have

$$\frac{dl}{ds} = \cos\gamma + \cos\beta(s)\frac{dt}{ds}.$$

From Corollary 3.7.1 and the Gauss–Bonnet theorem follows $\beta(s) = \pi - \alpha - \gamma + \bar{o}(s)$. Therefore,

$$\frac{dl}{ds} = \cos\gamma - \cos(\alpha+\gamma)\frac{dt}{ds} + \bar{o}(s).$$

Integrating the last equality with respect to s from 0 to δ, we obtain

$$l(\delta) = BC = \cos\gamma \cdot \delta - \cos(\alpha+\gamma)AB + \bar{o}(\delta).$$

Analogously, we obtain

$$AB = \cos\alpha \cdot \delta - \cos(\alpha+\gamma)BC + \bar{o}(\delta).$$

So,

$$BC + \cos(\alpha+\gamma)AB = \cos\gamma \cdot \delta + \bar{o}(\delta),$$
$$BC \cos(\alpha+\gamma) + AB = \cos\alpha \cdot \delta + \bar{o}(\delta).$$

From these equalities follows

$$AB = \frac{\cos\alpha - \cos\gamma\cos(\alpha+\gamma)}{\sin^2(\alpha+\gamma)}\delta + \bar{o}(\delta) = \frac{\sin\gamma \cdot \delta}{\sin(\alpha+\gamma)} + \bar{o}(\delta),$$

$$BC = \frac{\cos\gamma - \cos\alpha\cos(\alpha+\gamma)}{\sin^2(\alpha+\gamma)}\delta + \bar{o}(\delta) = \frac{\sin\alpha \cdot \delta}{\sin(\alpha+\gamma)} + \bar{o}(\delta).$$

Finally, the formula $\alpha + \beta + \gamma = \pi + \bar{o}(\delta)$ follows from (3.81). $\qquad\square$

Consider a simply connected complete surface of nonpositive Gaussian curvature. Denote such a surface by Φ_-. Recall that a surface Φ is *simply connected* if any simple closed curve divides Φ into two regions, one of which is homeomorphic to a disk.

Theorem 3.7.2. *There exists no more than one geodesic through any two points on a surface* Φ_-.

Proof. Assume the contrary. Let two different geodesics γ_1 and γ_2 pass through the points P_1 and P_2. Let Q_1 be the first point of intersection of γ_1 with γ_2 counting from P_1. Consider a piecewise smooth curve (a *bi-angle*) γ, composed of the arcs γ_1, γ_2 that join P_1 and Q_1. Let D be homeomorphic to the disk region bounded by γ. Apply the Gauss–Bonnet formula to D:

$$\iint_D K \, dS + \pi - \alpha + \pi - \beta = 2\pi, \tag{3.84}$$

where α and β are the angles at the vertices P_1 and Q_1, respectively. From (3.84) we obtain

$$\alpha + \beta = \iint_D K \, dS \leq 0,$$

which is impossible. $\qquad\square$

Corollary 3.7.3. There are no closed geodesics on a surface Φ_-.

Some other theorems can be deduced from Theorem 3.7.2.

Theorem 3.7.3. *Any two points on* Φ_- *can be joined by a unique shortest path.*

Proof. The existence of a shortest path follows from the completeness of Φ_- (see Theorem 3.5.6), and the uniqueness follows from Theorem 3.7.2. $\qquad\square$

Theorem 3.7.4. *Any arc of a geodesic on* Φ_- *is a shortest path.*

Proof. Indeed, if an arc PQ of some geodesic γ is not a shortest path, then joining its endpoints by a shortest path, we obtain two different geodesics joining two points, contrary to Theorem 3.7.2. $\qquad\square$

3.8 Local Comparison Theorems for Triangles

If a *geodesic triangle* (i.e., composed of geodesics) lies on a *convex* surface (i.e., on a surface of nonnegative Gaussian curvature), then the sum of its inner angles is at least π, which follows from (3.80). This property of geodesic triangle can be determined more exactly in the following sense. Let $\triangle ABC$ be a triangle on a complete *convex* surface Φ, composed of the shortest paths AB, AC, and BC. The *angle of the triangle* $\triangle ABC$ at some vertex is the angle between the shortest paths starting at this vertex. Thus the angles of any triangle $\triangle ABC$ are not greater than π.

We introduce standard notation: denote by α, β and γ the angles at the vertices A, B, and C, respectively. The lengths of the shortest paths AB, BC, and AC will be denoted by the same symbols. Note that $\triangle ABC$ is not uniquely defined by its vertices. Take on the plane $\mathbb{R}^2$ a triangle $\triangle A'B'C'$ whose sides are $A'B' = AB$,

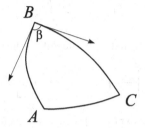

Figure 3.8. A small (geodesic) triangle on a surface.

$A'C' = AC$, $B'C' = BC$. The triangle $\triangle A'B'C'$ is called a *comparison triangle* for $\triangle ABC$, and its angles are denoted by α', β', and γ'. Then it turns out that *each angle of $\triangle ABC$ is not smaller than the corresponding angle of $\triangle A'B'C'$*.

In this section we prove the last statement for "small" triangles. We first prove

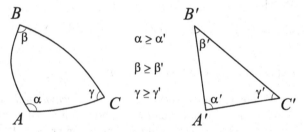

Figure 3.9. Comparison triangle on a plane for a small triangle on a convex surface.

an auxiliary comparison lemma on solutions $y(x)$ and $z(x)$ of the differential equations

$$y'' + k_1(x)y = 0, \qquad z'' + k_2(x)z = 0 \tag{3.85}$$

with initial conditions

$$y(0) = z(0) = 0, \qquad y'(0) = z'(0) = 1. \tag{3.86}$$

Lemma 3.8.1. If $k_1(x) \geq k_2(x)$ and $y(x) > 0$ for $x \in (0, x_0)$, then $z(x)/y(x) \geq 1$ is an increasing function on the interval $(0, x_0)$.

Remark 3.8.1. For the multidimensional case, the statement analogous to Lemma 3.8.1 is called *Rauch's comparison theorem*; see [Kl3].

Proof. Multiply (3.85:a) by $z(x)$, and (3.85:b) by $y(x)$, and then take their difference. We obtain

$$y''z - z''y + (k_1 - k_2)yz = 0. \qquad (3.87)$$

Integrating (3.87) from 0 to x, we obtain

$$y'(x)z(x) - z'(x)y(x) = \int_0^x (k_2 - k_1)yzdx.$$

Divide the last equality by zy and then integrate from 0 to x. We have

$$\log\frac{y}{z} = \int_0^x \left[\frac{1}{yz}\int_0^x (k_2 - k_1)yzdx\right]dx = -h(x). \qquad (3.88)$$

From (3.88) follows $\frac{y(x)}{z(x)} = \exp(h(x))$. The function $h(x)$, in view of the conditions of the lemma, is increasing, and $h(x) > 0$ for $x > 0$. □

From Lemma 3.8.1 some corollaries follow by a natural way.

Corollary 3.8.1. The inequality $y(x) \le z(x)$ is satisfied for $x \in (0, x_0]$.

Corollary 3.8.2. If $x_1 \in (0, x_0]$ and $y(x_1) = z(x_1)$, then $k_1(x) = k_2(x)$, $y(x) = z(x)$ for $x \in (0, x_1]$.

Corollary 3.8.3. If $z(x_0) = 1$, then $y(x) = z(x)$ and $k_1(x) = k_2(x)$ for $x \in (0, x_0]$.

Now we give the definition of a *"small"* triangle.

Definition 3.8.1. The real number $r_i(P)$ is called the *injectivity radius* of a complete surface Φ *at a point P* if any arc of a geodesic with initial point P whose length l is smaller than $r_i(P)$ is a shortest path and for any $r > r_i(P)$ there is an arc of a geodesic with the length r and initial point P that is not a shortest path. If the real number $r_i = \inf\{r_i(P): P \in \Phi\}$ differs from zero, then it is called the *injectivity radius of a surface* Φ.

From the definition of $r_i(P)$ and Theorem 3.5.6 it follows that the exponential map on an open disk $K(r_i(P))$ in the plane $T\Phi_P$ with center at P and radius $r_i(P)$ is a bijection of $K(r_i(P))$ onto a region $W(P) = \exp_P(K(r_i(P)))$ of Φ.

Definition 3.8.2. A triangle $\triangle ABC$ is called *admissible for the vertex A* if the distance from A to any point on the side BC is not greater than $r_i(P)$.

Take a triangle $\triangle\bar{A}\bar{B}\bar{C}$ on $\mathbb{R}^2$ whose sides are $\bar{A}\bar{B} = AB$, $\bar{A}\bar{C} = AC$, and $\angle\bar{B}\bar{A}\bar{C} = \alpha$.

Theorem 3.8.1. *If a triangle $\triangle ABC$ on a convex complete surface Φ of class C^2 is admissible for the vertex A, then $BC \le \bar{B}\bar{C}$.*

Proof. Take two rays on the plane $T\Phi_A$ tangent to the shortest paths AC and AB, and then mark off along them the line segments $A\tilde{C}$ and $A\tilde{B}$ whose lengths are equal to AC and AB, respectively. The triangle $\triangle A\tilde{B}\tilde{C}$ is equal to $\triangle \bar{A}\bar{B}\bar{C}$. In particular,

$$\tilde{B}\tilde{C} = \bar{B}\bar{C}. \tag{3.89}$$

Define geodesic polar coordinates $(\tilde{\rho}, \tilde{\varphi})$ and (ρ, φ) in the disk $K(r_i(A))$ with center A and radius $r_i(A)$ in the plane $T\Phi_A$ and in the region $W(A) = \exp_A K(r_i(A))$, respectively. Let the equations of the line segment $\tilde{B}\tilde{C}$ in coordinates $(\tilde{\rho}, \tilde{\varphi})$ be written in parametric form:

$$\tilde{\rho} = h(t), \qquad \tilde{\varphi} = \psi(t) \qquad (0 \le t \le 1).$$

Set $\gamma = \exp_A(\tilde{B}\tilde{C})$. Then the equations of γ in coordinates (ρ, φ) are given by the same functions $\rho = h(t), \varphi = \psi(t)$. Compare the length $l(\gamma)$ of γ with the length of the line segment $\tilde{B}\tilde{C}$:

$$\tilde{B}\tilde{C} = \int_0^1 \sqrt{[h'(t)]^2 + h^2(t)[\psi'(t)]^2}\, dt, \tag{3.90}$$

$$l(\gamma) = \int_0^1 \sqrt{[h'(t)]^2 + f^2(h(t), \psi(t))[\psi'(t)]^2}\, dt, \tag{3.91}$$

where the function $f(\rho, \varphi)$ satisfies (3.85:a) with the initial conditions $f(0, \varphi) = 0$, $f'_\rho(0, \varphi) = 1$. Then from Lemma 3.8.1 and the conditions of the theorem follows

$$f(h(t), \varphi(t)) \le h(t). \tag{3.92}$$

From (3.90)–(3.92) follows $l(\gamma) \le \tilde{B}\tilde{C}$, but the length of the shortest path BC is not greater than $l(\gamma)$. Hence we have $BC \le l(\gamma) \le \tilde{B}\tilde{C} = \bar{B}\bar{C}$. □

From Theorem 3.8.1 we obtain the comparison theorem for the angles of admissible triangles.

Theorem 3.8.2. *Under the conditions of Theorem 3.8.1 the inequality $\alpha \ge \alpha'$ holds, where α' is the angle of the comparison triangle $\triangle A'B'C'$ at the vertex A'.*

Proof. Since $BC = B'C' \le \tilde{B}\tilde{C}$, then by a well-known theorem of elementary geometry, $\alpha' \le \alpha$. □

Corollary 3.8.4. If a triangle $\triangle ABC$ on a complete convex surface Φ is admissible for each of its vertices, then the inequalities $\alpha \ge \alpha'$, $\beta \ge \beta'$, $\gamma \ge \gamma'$ hold.

The statements analogous to Theorems 3.8.1, 3.8.2 (with opposite signs in the inequalities) are true for saddle surfaces.

Problem 3.8.1. If $\triangle ABC$ on a complete saddle surface Φ of class C^2 is admissible for its vertex A, then the angle α is not greater than the angle α' of a comparison triangle $\triangle A'B'C'$.

Hint. It is sufficient to show that $BC \geq \bar{B}\bar{C}$ (under the notation of Theorem 3.8.1). For this it is sufficient to compare the length of the shortest path BC with the length of the curve $\bar{\gamma} = \exp_A^{-1}(BC)$ and then to apply Lemma 3.8.1.$\square$

Problem 3.8.2. The angles of an arbitrary triangle $\triangle ABC$ on a complete simply connected saddle surface Φ of class C^2 are not greater than the corresponding angles of a comparison triangle $\triangle A'B'C'$; $\alpha \leq \alpha', \beta \leq \beta', \gamma \leq \gamma'$.

If the Gaussian curvature of some surface Φ is not smaller than some real number k_0, then it is natural to compare the angles of $\triangle ABC$ on Φ with the angles of a comparison triangle on a *plane* $\mathbb{R}_{k_0}$ of constant curvature k_0. Denote by $(\triangle A'B'C')_{k_0}$ a *comparison triangle on* $\mathbb{R}_{k_0}$, and a surface Φ itself whose Gaussian curvature is not smaller than k_0, denote by Φ_{k_0}.

If $k_0 < 0$, then $\mathbb{R}_{k_0}$ is actually a *Lobachevski (hyperbolic) plane*, and if $k_0 > 0$, then $\mathbb{R}_{k_0}$ is a sphere of radius $1/\sqrt{k_0}$. In the last case the comparison triangle $(\triangle A'B'C')_{k_0}$ exists if and only if the perimeter of $\triangle ABC$ is not greater than $2\pi/\sqrt{k_0}$. Repeating the proof of Theorems 3.8.1 almost word for word and 3.8.2, we can deduce the following Theorem 3.8.3.

Theorem 3.8.3. *If a triangle $\triangle ABC$ on Φ_{k_0} is admissible for all its vertices, then its angles are not smaller than the corresponding angles of a comparison triangle $(\triangle A'B'C')_{k_0}$. If $k_0 > 0$, then the perimeter of $\triangle ABC$ is assumed not greater than $2\pi/\sqrt{k_0}$.*

Remark 3.8.2 (of the editor). Let (X, d_X) be a metric space. For a continuous path $\gamma: [0, L] \to X$, we define the *length* $l(\gamma)$ of γ by $l(\gamma) = \sup\{\sum_{i=0}^{n-1} d_X(\gamma(t_i), \gamma(t_{i+1})): 0 = t_0 < \cdots < t_n = L\}$, where the supremum is taken over all sequences $\{t_i\}_{0 \leq i \leq n}$ as above, and all $n \in \mathbb{N}$. A continuous path $\gamma: [0, L] \to X$ is a *geodesic* if it has a constant speed and is locally minimizing, that is,

$$l(\gamma|_{[a,b]}) = (|b - a|/L) \cdot l(\gamma),$$

and if for every $a \in [0, L]$, there is some $\varepsilon > 0$ such that

$$l(\gamma|_{[a',a'']}) = d_X(\gamma(a'), \gamma(a''))$$

holds, where $a' = \max\{a - \varepsilon, 0\}$ and $a'' = \min\{a + \varepsilon, L\}$.

A metric space (X, d_X) is called a *CAT(k)-space* if it satisfies the following conditions:

(i) *Every two points $x, y \in X$ (with $d_X(x, y) \leq \pi/\sqrt{k}$ if $k > 0$) are joined by a minimizing geodesic; that is, $\gamma: [0, L] \to X$, which satisfies $\gamma(0) = x, \gamma(L) = y$, and $l(\gamma) = d_X(x, y)$.*

(ii) (CAT(k)-property). *For an arbitrary geodesic triangle $\Delta(A, B, C) \subset X$ (with perimeter $< 2\pi/\sqrt{k}$ if $k > 0$) we have the comparison triangle $\Delta(A', B', C') \subset \mathbb{R}_k$ (with the same side lengths as $\Delta(A, B, C)$) such that $d_X(x, y) \leq d_{\mathbb{R}_k}(x', y')$ for every pair $x \in AB$, $y \in AC$ and the corresponding points $x' \in A'B'$, $y' \in A'C'$.*

The notion of CAT(k)-spaces (named for Cartan, Aleksandrov, and Toponogov), introduced by M. Gromov,[14] is based on Aleksandrov's original notion, i.e., *spaces with curvature bounded above by* $k \in \mathbb{R}$. Two-dimensional examples of CAT(1)-spaces are spheres of radius ≥ 1. More generally, complete smooth surfaces with Gaussian curvature uniformly bounded above by 1 and of injectivity radii bounded below by π are CAT(1)-spaces. Examples of CAT(0)-spaces are saddle surfaces in $\mathbb{R}^3$ and generalizations of *Hadamard manifolds*, which are simply connected complete Riemannian manifolds such that the sectional curvature is nonpositive.

A basic property of CAT(k)-spaces is that between any two geodesic segments AB and AC in X starting from one point there is an angle $\angle BAC$.

(iii) (Angle comparison theorem). *The angles α, β, γ of an arbitrary triangle T in X are not greater than the corresponding angles α', β', γ' of the comparison triangle T_k on $\mathbb{R}_k$.*

According to Reshetnyak's gluing lemma,[15] the space constructed by gluing CAT(k)-spaces isometrically along proper convex subsets is again a CAT(k)-space.

3.9 Aleksandrov Comparison Theorem for the Angles of a Triangle

The following comparison theorem for the angles of triangles by A.D. Aleksandrov holds (see proof below, in Section 3.9.1).

Theorem 3.9.1. *The angles of a triangle $\triangle ABC$ on a surface Φ_{k_0} of class C^2 are not smaller than the corresponding angles of a comparison triangle $(\triangle A'B'C')_{k_0}$:* $\alpha \geq \alpha'$, $\beta \geq \beta'$, $\gamma \geq \gamma'$.

First we shall prove three lemmas.

Lemma 3.9.1 (about convex quadrilaterals, A.D. Aleksandrov). Let $ABCD$ and $A'B'C'D'$ be two given convex quadrilaterals in the plane $\mathbb{R}_{k_0}$ whose corresponding sides are equal: $AB = A'B'$, $BC = B'C'$, $CD = C'D'$, $DA = D'A'$.

---

[14] M. Gromov, *Geometric Group Theory, Essays in Group Theory* (S.M. Gersten, ed.), M.S.R.I. Publ. 8, Springer-Verlag, Berlin-Heidelberg-New York, 1987, 75–264.

[15] Yu. G. Reshetnyak, *On the theory of spaces of curvature not greater than K*, Mat. Sb., 52 (1960), 789–798.

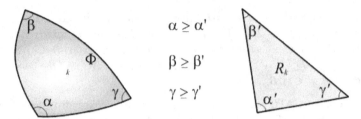

Figure 3.10. Comparison triangle $(\triangle A'B'C')_{k_0}$ for a triangle on a surface.

Then if the angle $\angle ADC$ is not smaller (greater) than $\angle A'D'C'$, then $\angle BAD$ and $\angle BCD$ are not greater (smaller) than $\angle B'A'D'$ and $\angle B'C'D'$, respectively. For the case $k_0 > 0$ we assume that the perimeter of $ABCD$ is smaller than $2\pi/\sqrt{k_0}$. The case that $\angle ADC$ is equal to π also is not excluded.

Proof. Let $k_0 = 0$. The conditions of the lemma implies that inequality $AC \geq A'C'$ ($AC > A'C'$), from which it follows that $\angle ABC$ is not smaller (greater) than $\angle A'B'C'$. But the sum of all angles of the quadrilaterals $ABCD$ and $A'B'C'D'$ is 2π. Therefore, at least one of the angles $\angle BAD$ or $\angle BCD$ does not exceed (is smaller than) $\angle B'A'D'$ or $\angle B'C'D'$. In both cases we have $BD \leq B'D'$ ($BD < B'D'$), whence the statement of the lemma follows.

   In the case of $k_0 \neq 0$ consider a quadrilateral $A''B''C''D''$ for which $\angle A''D''C'' = \angle A'D'C'$, $\angle B''C''D'' = \angle BCD$, and the sides $A''D''$, $D''C''$ and $C''B''$ are equal, respectively, to AD, DC, and CB. Comparing this quadrilateral with $ABCD$ and $A'B'C'D'$, one can see that $\angle BCD$ does not exceed (is smaller than) $\angle B'C'D'$.[16] Similarly for the second inequality, whence it follows that $\angle BAD$ does not exceed (is smaller than) $\angle B'A'D'$.   □

Remark 3.9.1. From the *law of cosines* of spherical and hyperbolic trigonometry ($k_0 = \pm 1$),

$$\cos a = \cos b \cos c + \sin a \sin b \cos A,$$
$$\cosh a = \cosh b \cosh c - \sinh a \sinh b \cos A$$

it follows that in all cases that $a = a(A)$ is a monotonic increasing function of the angle $A \in [0, \pi]$ when b, c are given. This simple fact on the monotonicity relation between angle and length of the closing edge of a geodesic hinge in $\mathbb{R}_{k_0}$ can be used in the proof of Lemma 3.9.1, case $k_0 \neq 0$. (A *geodesic hinge* in Φ is a figure consisting of a point $A \in \Phi$ called the *vertex* and minimal geodesics γ, τ emanating from A called *sides*. The angle between the tangent vectors to γ, τ at A is called the angle of the geodesic hinge.)

---

[16] $\triangle B''C''D'' = \triangle BCD$; hence $B''D'' = BD$ and $\angle A''D''B'' \geq \angle ADB$. Comparing $\triangle A''D''B''$ with $\triangle ADB$, we get $A''B'' \leq AB$ (see Remark 3.9.1 by editor). Comparing $\triangle A''B''C''$ with $\triangle A'B'C'$, from $A'B' \geq A''B''$ we get $\angle A'C'B' \geq \angle A''C''B''$. Then $\angle D''C''B'' \leq \angle D'C'B'$. Comparing $\triangle B'C'D'$ with $\triangle BCD$, we get $D'B' \geq DB$.

Let D be an open region on Φ whose closure is compact, with a triangle $\triangle ABC$ in its interior, and d an *elementary length of D* (see Theorem 3.5.4).

Definition 3.9.1. A triangle $\triangle ABC$ is a *thin triangle* if the distance from any point on the side AB up to the side AC does not exceed $\delta = d/4$.

Lemma 3.9.2. (about thin triangles.) If a triangle $\triangle ABC$ on a surface Φ_{k_0} is thin, then its angles at the vertices B and C are not smaller than the corresponding angles of a comparison triangle $(\triangle A'B'C')_{k_0}$: $\beta \geq \beta'$, $\gamma \geq \gamma'$. In the case $k_0 > 0$, the perimeter of $\triangle ABC$ is supposed to be not greater than $2\pi/\sqrt{k_0}$.

Proof. Introduce the parameterizations $\overset{\bullet}{B}(x)$ and $C(y)$ on the shortest paths AB and AC, where x and y are the lengths of the arcs $AB(x)$ and $AC(y)$ of the shortest paths AB and AC. Denote by $\beta(x,y)$ and $\gamma(x,y)$ the angles of $\triangle AB(x)C(y)$ at the vertices $B(x)$ and $C(y)$, respectively. Define the set T of ordered pairs (x,y) by the following conditions:

(1) The angles $\beta(x,y)$ and $\gamma(x,y)$ are not smaller than the corresponding angles $\beta'(x,y)$ and $\gamma'(x,y)$ of a comparison triangle $(\triangle A'B'C')_{k_1}$, where k_1 is an arbitrary real number smaller than k_0.
(2) $|x - y| \leq \delta/2$.
(3) If a pair (x_1, y_1) is in T, then (x_2, y_2) is also in T when $x_2 \leq x_1$, $y_2 \leq y_1$, and $|x_2 - y_2| \leq \delta/2$.

From Corollary 3.8.4 it follows that the pairs (x, y) for $x < \delta/2$ and $y < \delta/2$ belong to T. Consequently, the set T is nonempty. Define on T the function $f(x, y) = x + y$. Let $T_0 = \max\{f(x, y): (x, y) \in T\}$. If $T_0 = AB + AC$, then the statement of the lemma has been proved. Assume that $T_0 = x_0 + y_0 < AB + AC$ and derive a contradiction. Let, for definiteness, $x_0 \geq y_0$. Then prove that the angle $\beta(x, y)$ is greater than the angle $\beta'(x, y)$. Indeed, all the angles of $\triangle B(x_0 - \delta/2)B(x_0)C(y_0)$ are greater than the corresponding angles of the triangle $(\triangle B'(x_0 - \delta/2)B'(x_0)C'(y_0))_{k_1}$ (see Corollary 3.8.4 and the condition for k_1). Therefore, the angle at the vertex $B'(x_0 - \delta/2)$ of the convex quadrilateral $A''B''(x_0 - \delta/2)B''(x_0)C''(y_0)$ in the plane $\mathbb{R}_{k_1}$ obtained by gluing triangles $(\triangle A'B'(x_0 - \delta/2)C'(y_0))_{k_1}$ and $(\triangle B(x_0 - \delta/2)B(x_0)C(y_0))_{k_1}$ to each other along their common side $B'(x_0 - \delta/2)B'(x_0)$ is smaller than π. Applying Lemma 3.9.1 to the quadrilateral $A''B''(x_0 - \delta/2)B''(x_0)C''(y_0)$ and the triangle $(\triangle A'B'(x_0)C'(y_0))_{k_1}$, we obtain our statement $\beta(x_0, y_0) > \beta'(x_0, y_0)$. From the last inequality and continuity follows the existence of $\delta_1 > 0$ such that for $0 \leq t \leq \delta_1$ the angle $\beta(x_0, y_0 + t)$ is not smaller than $\beta'(x_0, y_0 + t)$.

We prove that the pair $(x_0, y_0 + t)$ for $0 \leq t \leq \min\{\delta_1, \delta, AC - y_0\} = \delta_2$ belongs to T. Indeed, all angles of $\triangle B(x_0)C(y_0 + t)C(y_0)$ are greater than the corresponding angles of the triangle $(\triangle B'(x_0)C'(x_0 + t)C'(y_0))_{k_1}$. Therefore, arguments similar to those stated above show us that the angle $\gamma(x_0, y_0 + t)$ is greater than $\gamma'(x_0, y_0 + t)$ for $0 \leq t \leq \delta_2$. But then $f(x_0, y_0 + t) = x_0 + y_0 + t > x_0 + y_0 = T_0$ for $t > 0$, contrary to the definition of T_0. The statement of the lemma now follows from the arbitrariness of k_1. $\qquad\square$

Lemma 3.9.3 (about a limit angle). Let the shortest paths AB, BC, CX_n be given, and $C \notin AB$, $X_n \neq B$, $AX_n < AB$. Denote by α the angle between BA and BC, by β_n the angle between X_nC and X_nA. If $\lim_{n \to \infty} X_n = B$, then $\lim_{n \to \infty} \beta_n = \beta$ and $\beta \leq \alpha$.

Proof. Obviously, it is sufficient to prove that if a sequence of shortest paths CX_n converges to some shortest path $\overline{CB}$, then $\beta \leq \alpha$. Suppose that $\beta > \alpha$ and obtain a contradiction. Draw through a point X_n a geodesic σ_n under the angle $\pi - \beta_n$ to the shortest path X_nB, so that σ_n intersects the shortest path CB at some point C_n, and from the point C draw a geodesic $\bar{\sigma}_n$ forming an angle α with the shortest path CX_n, so that it intersects the shortest path CX_n at some point $\bar{C}_n$. For sufficiently

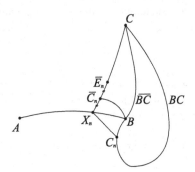

Figure 3.11. A limit angle.

large n, the existence of geodesics σ_n and $\bar{\sigma}_n$ with the above-mentioned properties results from the assumption $\beta > \alpha$. From the statement of Problem 3.7.2 we obtain the equalities

$$\angle BC_nX_n = \angle B\bar{C}_nX_n = \beta_n - \alpha + \bar{o}(BX_n),$$

$$BC_n = B\bar{C}_n = \frac{\sin \beta_n}{\sin(\beta_n - \alpha)} BX_n + \bar{o}(BX_n), \tag{3.93}$$

$$C_nX_n = \bar{C}_nX_n = \frac{\sin \alpha}{\sin(\beta_n - \alpha)} BX_n + \bar{o}(BX_n).$$

Furthermore, from the triangle inequality we obtain

$$B\bar{C}_n + \bar{C}_nC \geq BC = CC_n + C_nB,$$
$$X_nC_n + C_nC \geq X_nC = X_n\bar{C}_n + C\bar{C}_n. \tag{3.94}$$

From (3.93) and (3.94) follows the equality

$$C\bar{C}_n = CC_n + \bar{o}(BX_n). \tag{3.95}$$

Take now a point $\bar{E}_n$ on the shortest path CX_n such that $\bar{C}_n\bar{E}_n = \bar{C}_nX_n$ and $\bar{E}_n \neq X_n$. Then from the statement of Problem 3.7.2 it is easy to deduce that

$$C\bar{E}_n = 2\bar{C}_n X_n \sin\left(\frac{\pi}{2} - \frac{\beta_n - \alpha}{2}\right) + \bar{o}(BX_n)$$

$$= 2\bar{C}_n X_n \cos\left(\frac{\beta_n - \alpha}{2}\right) + \bar{o}(BX_n).$$

From this equality and from (3.95) follows

$$CC_n + C_n B = CB \leq C\bar{E}_n + \bar{E}_n C$$

$$= 2\bar{C}_n X_n \cos\left(\frac{\beta_n - \alpha}{2}\right) + \bar{E}_n C + \bar{o}(BX_n)$$

$$= 2\bar{C}_n X_n \cos\left(\frac{\beta_n - \alpha}{2}\right) + C\bar{C}_n - C_n X_n + \bar{o}(BX_n). \qquad (3.96)$$

Further on, from (3.93), (3.95), and (3.96) we obtain the inequality

$$2CC_n \leq 2\bar{C}_n X_n \cos\left(\frac{\beta_n - \alpha}{2}\right) + \bar{o}(BX_n)$$

$$= 2C\bar{C}_n \cos\left(\frac{\beta_n - \alpha}{2}\right) + \bar{o}(BX_n). \qquad (3.97)$$

Divide (3.97) by $2CC_n$ and pass to the limit for $n \to \infty$. Then we obtain $1 \leq \cos\left(\frac{\beta - \alpha}{2}\right)$, which is impossible, since $\beta - \alpha > 0$. $\qquad \square$

3.9.1 Proof of the Comparison Theorem for the Angles of a Triangle

Let $\triangle ABC$ be an arbitrary triangle on a surface Φ_{k_0}. In the case of $k_0 > 0$ we assume temporarily that the perimeter of $\triangle ABC$ is smaller than $2\pi/\sqrt{k_0}$. We prove the theorem for the angle α. Introduce on AC a parameterization $C(x)$, where x is the length of an arc $AC(x)$ of shortest path. Denote by $\gamma(x)$ the angle $\angle AC(x)B$ of the triangle $\triangle ABC$. Define the set of real numbers T of x satisfying the following inequalities:

$$\alpha \geq \alpha', \qquad \gamma \geq \gamma' \qquad (3.98)$$

The set T is not empty, since by virtue of the lemma about *thin triangles* and the lemma about a *limit angle*, the inequalities (3.98) are satisfied for sufficiently small x. Let $x_0 = \sup T$. If $x_0 = AC$, then Theorem 3.8.1 is proven for the angle α.

Assume $x_0 < AC$, and obtain a contradiction. Note that $x_0 \in T$ by the lemma about a *limit angle*. Take a sequence of points $C_n = C(x_n)$, $C_n \neq C(x_0) = C_0$, $x_n > x_0$ such that $\lim_{n\to\infty} C_n = C_0$. Also assume, without loss of generality, that the sequence of the shortest paths BC_n converges to some shortest path BC_0. For sufficiently large n ($n \geq n_0$) the triangle $\triangle C_{n_0} BC_0$ is thin. Therefore, the angle at the vertex C_0'' of the quadrilateral $A''B''C_{n_0}''C_0''$ obtained by gluing $(\triangle A'B'C_{n_0}')_{k_0}$ and $(\triangle C_{n_0}'B'C_0')_{k_0}$ along their common side $B'C_0'$ does not exceed π. Hence, from Lemma 3.9.1 it follows that $x_{n_0} \in T$, contrary to the definition of x_0. So, Theorem 3.9.1 is proven for the angle α. For the other angles of $\triangle ABC$, Theorem 3.9.1 can be proved similarly.

We are left to consider the case of $k_0 > 0$ and remove from the assumption that the perimeter of $\triangle ABC$ is smaller than $2\pi/\sqrt{k_0}$. Assume that on the surface Φ_{k_0} there is a triangle $\triangle ABC$ whose perimeter is greater than $2\pi/\sqrt{k_0}$, and obtain a contradiction. Suppose that at least one of the angles of $\triangle ABC$, say the vertex A, differs from π. Take the points B_0 and C_0 on the sides AB and AC such that $AB_0 + B_0C_0 + C_0A = 2\pi/\sqrt{k_0}$. Assume, without loss of generality, that the sum of the lengths of the two smallest sides of $\triangle AB_0C_0$ is equal to the length of the third. Let B_n and C_n be the sequences of points on AB and AC; moreover, $B_n \neq B_0$, $C_n \neq C_0$, $AB_n < AB_0$, $AC_n < AC_0$, and $\lim_{n\to\infty} B_n = B_0$, $\lim_{n\to\infty} C_n = C_0$. Then for every t it is true that $AB_n + B_nC_n + C_nA < 2\pi/\sqrt{k_0}$, for otherwise, the perimeter of $\triangle ABC$ would be $2\pi/\sqrt{k_0}$. Applying Theorem 3.9.1 to triangles $\triangle AB_nC_n$, we obtain $\beta \geq \beta'$, $\gamma \geq \gamma'$. But as is easy to see, $\lim_{n\to\infty} \beta'_n = \lim_{n\to\infty} \gamma'_n = \pi$, and hence $\lim_{n\to\infty} \beta_n = \lim_{n\to\infty} \gamma_n = \pi$. But then B_0BCC_0 would be the shortest path of length $2\pi/\sqrt{k_0}$, and consequently, the perimeter of $\triangle ABC$ would be $2\pi/\sqrt{k_0}$, contrary to the assumption.

So we have proved that if the perimeter of $\triangle ABC$ is greater than $2\pi/\sqrt{k_0}$, then all its angles are π, which means that the line $AB \cup BC \cup CA$ is a closed geodesic γ. But then the perimeter of $\triangle A_1BC$, where $A_1 \in \gamma$ and A_1 also is close to A, is larger than $2\pi/\sqrt{k_0}$, and its angles at the vertices B and C differ from π. The obtained contradiction proves an absence on Φ_{k_0} of a triangle whose perimeter is greater than $2\pi/\sqrt{k_0}$.

Now let the perimeter of $\triangle ABC$ be $2\pi/\sqrt{k_0}$. If the triangle is nondegenerate, then reasoning as above, we can prove that $\triangle ABC$ is a closed geodesic, and consequently, all its angles are π, and then the statements of Theorem 3.9.1 are obviously true. If it is degenerate, that is, composed of two shortest paths (a *biangle* (or *lune*)), then in this case the surface Φ_{k_0} is a sphere (see Problem 3.10.1), and $\triangle ABC$ can be compared with itself.

The comparison theorem for the angles can formulated in a different form, using the convexity condition of A.D. Aleksandrov. Let AB and AC be two shortest paths starting from the point A, and also suppose $B(x) \in AB$, $C(y) \in AC$, $x = AB(x)$, $y = AC(y)$. Denote by $\varphi(x, y)$ the angle at the vertex A' of a comparison triangle $(\triangle A'B'(x)C'(y))_{k_0}$. We say that the *Aleksandrov convexity condition* is satisfied on a surface Φ with respect to a plane $\mathbb{R}_{k_0}$ if the function $\varphi(x, y)$ is decreasing.

Theorem 3.9.2. *A surface Φ_{k_0} satisfies the Aleksandrov convexity condition with respect to $\mathbb{R}_{k_0}$.*

Proof. This theorem easily follows from Theorem 3.9.1 and Lemma 3.9.1. $\square$

3.10 Problems to Chapter 3

Problem 3.10.1. The diameter d of a surface Φ_{k_0} ($k_0 > 0$) is not greater than $\pi/\sqrt{k_0}$. If $d = \pi/\sqrt{k_0}$, then Φ_{k_0} coincides with a sphere of radius $1/\sqrt{k_0}$.

Solution. The first statement of Problem 3.10.1 obviously follows from Theorem 3.9.1. Consider the case $d = \pi/\sqrt{k_0}$. Let A and B be the endpoints of a diameter, and P an arbitrary point on Φ_{k_0}. Then, by the triangle inequality, $AP + PB \geq \pi/\sqrt{k_0}$, but on the other hand, $\pi/\sqrt{k_0} + AP + PB \leq 2\pi/\sqrt{k_0}$. Therefore, $AP + PB = \pi/\sqrt{k_0}$. From the last equality it follows that the polygonal line $AP \cup PB$ is a shortest path of length $\pi/\sqrt{k_0}$. Thus we have obtained that any geodesic starting from A comes to B, and the length of an arc AB of this geodesic is $\pi/\sqrt{k_0}$. Introduce the geodesic polar coordinates (ρ, φ) with the origin at A. Then $ds^2 = d\rho^2 + f^2(\rho, \varphi)d\varphi^2$, where the function $f(\rho, \varphi)$ satisfies the equation $f''_{\rho\rho} + K(\rho, \varphi)f = 0$ with initial conditions $f(0, \varphi) = 0$, $f'_\rho(0, \varphi) = 1$. From what we have proved it follows that $f(\rho, \varphi) > 0$ for $0 < \rho < \pi/\sqrt{k_0}$, and since $\sin(\sqrt{k_0}\rho)/\sqrt{k_0}$ is zero at $\rho = \pi/\sqrt{k_0}$, then from the third corollary of Lemma 3.8.1 it follows that $K(\rho, \varphi) \equiv k_0$ for $0 \leq \varphi \leq 2\pi$; i.e., we have proved that the Gaussian curvature at each point of the surface Φ_{k_0} is the constant k_0. The statement of Problem 3.10.1 now follows from Theorem 2.8.2 of Liebmann. □

Remark 3.10.1. In the conditions of Problem 3.10.1 it is possible to construct an isometry $\psi \colon \Phi_{k_0} \to S(1/\sqrt{k_0})$ and not use Liebmann's theorem. So this problem is solved for the multidimensional case. The reader is asked to prove that $\exp_A$ is the required isometry ψ.

Recall that a *straight line on a surface* is a complete geodesic γ such that every one of its arcs is a shortest path.

Problem 3.10.2 (S. Cohn-Vossen). Prove that if on a complete convex surface of class C^2 there is a straight line γ, then Φ is a cylinder.

Solution. Take an arbitrary point $P \in \Phi$, $P \in \gamma$. Let γ_P be a point on γ, the nearest to P. Then the shortest path $P\gamma_P$ is either orthogonal to γ, or P can be joined with a point γ_P at least by two shortest paths, each of which forms with γ an angle not greater than $\pi/2$ (see Lemma 3.5.1). Let $\gamma(t)$ be a parameterization of γ and $t = \pm\gamma_P\gamma(t)$, $\gamma_P = \gamma(0)$, $-\infty < t < \infty$, t_n be a sequence of positive numbers tending to infinity, and τ_n a sequence of negative numbers tending to minus infinity. Without loss of generality, assume that the limit of the shortest paths $P\gamma(t_n)$ for $n \to \infty$ is some ray σ_1 with vertex at P, and the limit of the shortest paths $P\gamma(\tau_n)$ for c is some ray σ_2 with the same vertex P. Place the triangles $\triangle P'\gamma'(t_n)\gamma'(\tau_n)$ on $\mathbb{R}^2$ so that the sides $\gamma'(t_n)\gamma'(\tau_n)$ lie on the same straight line a. From Theorem 3.9.1 and Lemma 3.9.1 it follows that the distance from the vertex P' of triangle $\triangle P'\gamma'(t_n)\gamma'(\tau_n)$ up to a straight line a does not exceed $P\gamma_P$ for any n. Therefore, the limit of the angle $\angle \gamma'(t_n)P'\gamma'(\tau_n)$ for $n \to \infty$ is π; i.e., the rays σ_1 and σ_2 lie on the same geodesic $\bar{\gamma}$. But then from

Theorem 3.9.1 it follows that the angle between σ_1 and σ_2 is also π, since the rays σ_1 and σ_2 lie on the same geodesic $\bar{\gamma}$. Furthermore, the sum of the angles $\angle P'\gamma'(0)\gamma'(t_n)$ and $\angle \gamma'(0)P'\gamma'(t_n)$ of $\triangle \gamma'(0)P'\gamma'(t_n)$ for $n \to \infty$ is equal to π, and since $\angle P'\gamma'(0)\gamma'(t_n)$ does not exceed $\pi/2$ (by Theorem 3.9.1) for any n, the limit of the angle $\angle \gamma'(0)P'\gamma'(t_n)$ for $n \to \infty$ is not smaller than $\pi/2$. Then again by Theorem 3.9.1, an angle between the ray σ_1 and the shortest path $P\gamma(0)$ is not smaller than $\pi/2$. It can be proved similarly that the angle between σ_2 and $P\gamma(0)$ is not smaller than $\pi/2$. But since their sum is π, we obtain that $P\gamma(0)$ intersects $\bar{\gamma}$ in a right angle. From here it follows that a shortest path $P\gamma(0)$ also intersects the geodesic γ in a right angle.

Now let $P_1 \in \bar{\gamma}$ and $P_1 \neq P$. Repeating all the previous constructions and reasoning, we obtain that the shortest path $P_1\gamma_{P_1}$ intersects the geodesics $\bar{\gamma}$ and γ also in a right angle. So, we have obtained that in the region D bounded by the quadrilateral $PP_1\gamma_{P_1}\gamma_P$, all internal angles are $\pi/2$. Applying to region D the Gauss–Bonnet formula, we obtain that the integral curvature of D is zero. But the Gaussian curvature of the surface Φ is nonnegative; consequently, it is identically zero at each point of D. In particular, the Gaussian curvature of Φ is zero at P. But the point P has been selected arbitrarily. Hence the Gaussian curvature of Φ is zero at every point. $\qquad\square$

Remark 3.10.2. In Problem 3.10.2 as well as in Problem 3.10.1 it is possible to construct an isometry ψ of Φ onto the plane $\mathbb{R}_0$ without referring to the Gauss–Bonnet theorem. Namely, these arguments solve this problem in the n-dimensional case. To construct a map ψ, one needs to introduce a semigeodesic coordinate system on Φ and $\mathbb{R}_0$ and to compare points with identical coordinates.

Problem 3.10.3. For a triangle $\triangle ABC$ on a surface Φ_{k_0} let the shortest path AB be a unique shortest path connecting points A and B, and let the angle γ be equal to the angle γ'. Prove that then all angles of $\triangle ABC$ are equal to the corresponding angles of $(\triangle ABC)_{k_0}$.

Hint. It is sufficient to prove that for any point P on the shortest path AC the angle $\angle BPC$ is equal to $\angle B'P'C'$ of triangle $(\triangle B'P'C')_{k_0}$, which follows easily from Theorem 3.9.1 and Lemma 3.9.1. After this, it is easy to prove that $\alpha = \alpha'$. The equality $\beta = \beta'$ is proved analogously.

Problem 3.10.4. Prove that in the conditions of Problem 3.10.3, $\triangle ABC$ bounds a region such that at every one of its points the Gaussian curvature is k_0.

Hint. Use the results of Problem 3.10.3, Lemma 3.5.1, and Lemma 3.8.1.

Problem 3.10.5. Let $\triangle ABC$ be composed of the shortest paths AB, BC, and an arc AC of a geodesic whose length is smaller than $AB + BC$. Prove that $\alpha \geq \alpha'$ and $\gamma \geq \gamma'$.

Hint. Divide a geodesic AC into a finite number of arcs, each of them a shortest path, and take advantage of Theorem 3.9.1 and Lemma 3.9.1.

Problem 3.10.6. Prove that on a surface Φ_{k_0} for $k_0 > 0$ each arc of a geodesic whose length is greater than $4\pi/\sqrt{k_0}$ has points of self-intersection.

Hint. Assume the opposite and with the help of Theorem 3.9.1 reduce this assumption to a contradiction.

Problem 3.10.7. For $\triangle ABC$ on a surface Φ_{k_0} with $k_0 > 0$ construct a triangle $(\triangle A''B''C'')_{k_0}$ whose angles are equal to the corresponding angles of $\triangle ABC$. Prove that then $AB \leq A''B''$, $AC \leq AC''$, $BC \leq B''C''$. Consider the case in which $AB = A''B''$ and the angles of $\triangle ABC$ at the vertices A and B are equal to the angles of $\triangle A''B''C''$ at the vertices A'' and B''.

Problem 3.10.8. Formulate and solve the problems for saddle surfaces analogous to Problems 3.10.3 and 3.10.4.

Problem 3.10.9. Let K_r be a disk of radius r with center at a point O on a convex surface Φ, and let AB be a chord of this disk. Prove that if $\angle OAB = \angle OBA = 45°$, then $AB \geq \sqrt{2}r$.

Problem 3.10.10. Let r be some ray with vertex at a point P on a complete convex surface Φ of class C^2. Introduce on r a parameterization $r(t)$, where t is the arc length parameter counted from P. Let $B(t) = \{Q \in \Phi : \rho(Q, r(t)) < t\}$. Prove that $D(r) = \Phi \setminus \bigcup_{t=0}^{\infty} B(t)$ is an absolutely convex set on Φ for any ray r.

Problem 3.10.11. Let Φ be a complete regular surface of class C^2 whose Gaussian curvature satisfies the inequality $\frac{1}{a^2} \leq K \leq 1$. Prove that there is a diffeomorphism φ of Φ onto a unit sphere S_1 such that for any points P and Q on Φ, the following inequalities satisfied:

$$\rho_1(\varphi(P), \varphi(Q)) \leq \rho_\Phi(P, Q) \leq a\rho_1(\varphi(P), \varphi(Q)).$$

Here ρ_1 is a metric on S_1, and ρ_Φ is a metric on Φ.

Remark 3.10.3. The construction of diffeomorphisms satisfying the first and second inequality separately is simple enough. But it is not known whether there exists a diffeomorphism φ satisfying both these inequalities.

Note that all theorems and problems of Sections 3.9, 3.10 can be formulated and proved for any geodesically convex region on a regular surface of class C^2.

References

[AlZ]   Aleksandrov A.D. and Zalgaller V.A. *Intrinsic geometry of surfaces.* Translations of Mathematical Monographs, v. 15. Providence, RI: American Mathematical Society. VI, 1967.

[Bl]   Blaschke W. and Leichtweiss K. *Elementare Differentialgeometrie.* Berlin-Heidelberg-New York: Springer-Verlag. X, 1973.

[Bus]   Busemann H. *Convex Surfaces.* Interscience Tracts in Pure and Applied Mathematics. 6. New York–London: Interscience Publishers, 1958.

[Cohn] Cohn-Vossen S.E. *Verbiegbarkeit der Flächen im Grossen.* Uspechi Mat. Nauk, vol. 1, 33–76, 1936.

[Gau]   Gauss C.F. *General Investigations of Curved Surfaces.* Hewlett, NY: Raven Press, 1965.

[Hop]   Hopf H. *Differential Geometry in the Large.* 2nd ed. Lecture Notes in Mathematics, Vol. 1000. Berlin etc.: Springer-Verlag, 1989.

[Kl1]   Klingenberg W. *Riemannian Geometry.* 2nd ed. de Gruyter Studies in Mathematics. Berlin: Walter de Gruyter, 1995.

[Kl2]   Klingenberg W. *A Course in Differential Geometry.* Corr. 2nd print. Graduate Texts in Mathematics, v. 51. New York–Heidelberg–Berlin: Springer-Verlag. XII, 1983.

[Kl3]   Gromoll D., Klingenberg W. and Meyer W. *Riemannsche Geometrie im Grossen.* Lecture Notes in Mathematics. 55. Berlin-Heidelberg-New York: Springer-Verlag, VI, 1975.

[Ku1]   Kutateladze S.S. (ed). *A.D. Alexandrov Selected Works. Part 2: Intrinsic Geometry of Convex Surfaces.* New York; London: Taylor & Francis, 2004.

[Miln] Milnor J. *Morse Theory.* Annals of Mathematics Studies. No. 51, Princeton, Princeton University Press, VI, 1963.

[Top]   Toponogov V.A. *Riemannian spaces having their curvature bounded be-low by a positive number.* Dokl. Akad. Nauk SSSR, vol. 120, 719–721, 1958.

[Pog]   Pogorelov A.V. *Extrinsic Geometry of Convex Surfaces.* Translations of Mathematical Monographs, Vol. 35. Providence, American Mathematical Society, VI, 1973.

[Sto]   Stoker, J.J. *Differential Geometry.* Pure and Applied Mathematics. Vol. XX. New York etc.: Wiley-Interscience, 1969.

Additional Bibliography:

Differential Geometry of Curves and Surfaces

[Am1]   Aminov Yu. *Differential Geometry and Topology of Curves.* Amsterdam: Gordon and Breach Science Publishers, 2000.

[BG]   Berger M. and Gostiaux, B. *Differential Geometry: Manifolds, Curves and Surfaces.* New York: Springer-Verlag, 1988.

[Blo]   Bloch E.D. *A First Course in Geometric Topology and Differential Geometry.* Boston: Birkhäuser, 1997.

[Cas]   Casey J. *Exploring Curvature.* Wiesbaden: Vieweg, 1996.

[doC1]   do Carmo M. *Differential Geometry of Curves and Surfaces.* Englewood Cliffs, NJ: Prentice Hall, 1976.

[Gra]   Gray A. *A Modern Differential Geometry of Curves and Surfaces.* Second Edition, CRC Press, Boca Raton 1998.

[Hen]   Henderson D.W. *Differential Geometry: A Geometric Introduction.* Upper Saddle River, NJ: Prentice Hall, 1998.

[Hic]   Hicks N.J. *Notes of Differential Geometry.* New York: Van Nostrand Reinhold Company,1971.

[Kre]   Kreyszig E. *Differential Geometry* (Mathematical Expositions No. 11), 2nd Edition. Toronto: University of Toronto Press, 1964.

[Mill]   Millman R.S. and Parker G.D. *Elements of Differential Geometry.* Englewood Cliffs, NJ: Prentice-Hall, 1977.

[Opr]   Oprea J. *Introduction to Differential Geometry and Its Applications.* Upper Saddle River: Prentice Hall, 1997.

[Pra]   Prakash N. *Differential Geometry: An Integrated Approach.* New Delhi: Tata McGraw-Hill, 1993.

[Pre]   Pressley A. *Elementary Differential Geometry.* Springer-Verlag, 2001.

[Rov1]   Rovenski V.Y. *Geometry of Curves and Surfaces with MAPLE.* Boston, Birkhäuser, 2000.

[Shi]   Shikin E.V. *Handbook and atlas of curves.* Boca Raton, FL: CRC Press. xiv, 1995.

[Tho]   Thorpe J.A. *Elementary Topics in Differential Geometry.* Undergraduate Texts in Mathematics. Corr. 4th printing, Springer-Verlag. 1994.

Riemannian Geometry

[Am]   Aminov Yu. *The Geometry of Submanifolds*. Amsterdam: Gordon and Breach Science Publishers, 2001.

[Ber]   Berge M. *A Panoramic View of Riemannian Geometry*. Berlin: Springer, 2003.

[BCO]   Berndt J., Console S. and Olmos C. *Submanifolds and Holonomy*. Chapman & Hall/CRC Research Notes in Mathematics 434. Boca Raton, FL: Chapman and Hall/CRC, 2003.

[Bes2]   Besse A.L. *Einstein Manifolds*. Springer, Berlin-Heidelberg-New York, 1987.

[BH]   Bridson M.R. and Haefliger A. *Metric Spaces of Non-Positive Curvature*. Fundamental Principles of Mathematical Sciences, 319. Springer-Verlag, 2001.

[BZ]   Burago Yu.D. and Zalgaller V.A. *Geometric Inequalities*. Transl. from the Russian by A. B. Sossinsky. Grundlehren der Mathematischen Wissenschaften, 285. Berlin etc.: Springer-Verlag. XIV, 1988.

[BBI]   Burago D., Burago Yu. and Ivanov S. *A Course in Metric Geometry*. Graduate Studies in Mathematics. Vol. 33. Providence, RI: AMS, 2001.

[Chen1]   Chen, Bang-yen *Geometry of Submanifolds*. Pure and Applied Mathematics. 22. New York: Marcel Dekker, Inc., 1973.

[Chen2]   Chen, Bang-yen *Geometry of Submanifolds and Its Applications*. Tokyo: Science University of Tokyo. III, 1981.

[Cao]   Cao J. *A Rapid Course in Modern Riemannian Geometry*. Ave Maria Press, 2004.

[doC2]   do Carmo M. *Riemannian Geometry*. Boston: Birkhäuser, 1992.

[Car1]   Cartan, E. *Geometry of Riemannian Spaces*. Transl. from the French by J. Glazebrook. Notes and appendices by R. Hermann. Brookline, Massachusetts: Math Sci Press. XIV, 1983.

[Car2]   Cartan, E. *Riemannian geometry in an orthogonal frame*. From lectures delivered by Elie Cartan at the Sorbonne 1926–27. With a preface to the Russian edition by S.P. Finikov. Translated from the 1960 Russian edition by V.V. Goldberg and with a foreword by S.S. Chern.

[Cha]   Chavel I., et al. *Riemannian Geometry: A Modern Introduction*. (Cambridge Tracts in Mathematics). Cambridge University Press. Cambridge, 1993.

[CE]   Cheeger, J. and Ebin, D. *Comparison Theorems in Riemannian Geometry*. Elsevier, 1975.

[DFN]   Dubrovin B.A., Fomenko A.T., Novikov S.P. *Modern Geometry: Methods and Applications* Robert G. Burns, trans., New York: Springer-Verlag, 1984.

[Eis]   Eisenhart L.P. *Riemannian Geometry*. Princeton University Press, Princton, N.J. 1997 (originally published in 1926).

[GHL]   Gallot S., Hulin D., Lafontaine J. *Riemannian Geometry*. Universitext. Springer-Verlag; 3rd edition. 2004.

[Jos]   Jost J. *Riemannian Geometry and Geometric Analysis*. Springer-Verlag; 3rd edition, 2001.

[KN]   Kobayashi S. and Nomizu S. *Foundations of Differential Geometry* I, II. Wiley-Interscience, New York; New Edition, 1996.

[Lan]   Lang S. *Fundamentals of Differential Geometry*. Vol. 191, Springer-Verlag, 1998.

[Lau]   Laugwitz D. *Differential and Riemannian Geometry*. F. Steinhardt, trans., New York: Academic Press, 1965.

[Lee]   Lee J. M. *Riemannian Manifolds: An Introduction to Curvature*. (Graduate Texts in Mathematics , Vol. 176). New York: Springer, 1997.

[Mey]   Meyer W.T. *Toponogov's Theorem and Applications*. Lecture Notes, College on Differential Geometry, Trieste, 1989. http://www.math. uni-muenster.de/math/u/meyer/publications/toponogov.html

[Mrg]   Morgan F. *Riemannian Geometry: A Beginner's Guide*. A.K. Peters Ltd, 2nd edition, 1998.

[O'Neill] O'Neill B. *Elementary Differential Geometry*. Academic Press, London–New York, 1966.

[OG]   Osserman, R., Gamkrelidze, R.V. (eds.) *Geometry V: Minimal surfaces*. Transl. from the Russian. Encyclopaedia of Mathematical Sciences. 90. Berlin: Springer, 1997.

[Pet]   Petersen P. *Riemannian Geometry*. (Graduate Texts in Mathematics, 171). Springer-Verlag, 1997.

[Pos]   Postnikov M.M. *Geometry VI · Riemannian Geometry*. Encyclopaedia of Mathematical Sciences. Vol. 91. Berlin: Springer, 2001.

[Rov2]   Rovenski V.Y. *Foliations on Riemannian Manifolds and Submanifolds. Synthetic Methods*. Birkhäuser, 1997.

[Sak]   Sakai T. *Riemannian Geometry*. American Mathematical Society, 1996.

[Spi]   Spivak M. *A Comprehensive Introduction to Differential Geometry* (5 Volumes). Publish or Perish, 1979.

[Ste]   Sternberg S. *Lectures on Differential Geometry* (2nd edition). Chelsea. New York, 1982.

[Wal]   Walschap G. *Metric Structures in Differential Geometry*. Series: Graduate Texts in Mathematics, Vol. 224. Birkhäuser, 2004.

[Wil]   William M.B. *An Introduction to Differentiable Manifolds and Riemannian Geometry*. 2. Udgave, Academic Press, 1986.

[Wilm]   Willmore T.J. *Riemannian Geometry*. Oxford University Press, 1997.

Index